DIFFERENTIAL GEOMETRY

K. L. Wardle

DOVER PUBLICATIONS, INC.
Mineola, New York

Bibliographical Note

This Dover edition, first published in 2008, is an unabridged republication of the work originally published in 1965 by Dover Publications, Inc., Mineola, New York.

Library of Congress Cataloging-in-Publication Data

Wardle, K. L. (Kenneth Lansdell)
Differential geometry / K.L. Wardle. — Dover ed.
p. cm.
"An unabridged republication of the work originally publlished in 1965 by Dover Publications, Inc., Mineola, New York"—T.p. verso.
Includes index.
ISBN-13: 978-0-486-46272-1
ISBN-10: 0-486-46272-2
1. Geometry, Differential. I. Title.

QA641.W35 2008
516.3'6—dc22

2007046746

Manufactured in the United States of America
Dover Publications, Inc., 31 East 2nd Street, Mineola, N.Y. 11501

Preface

THE PURPOSE of this book is to give an elementary account of the differential geometry of curves and surfaces in Euclidean space of three dimensions. The development of the theory is by vector methods throughout, and it is assumed that the reader has already made some acquaintance with elementary vector results; as a reminder a list of these is included at the start.

The first chapter, on plane curves, is brief, its purpose merely being to introduce in a familiar context some of the ideas and techniques used later in the text.

In the second chapter an intuitive definition of the arc-length of a space curve has been given, and for a rigorous treatment the reader is asked to refer to books on analysis. The main object of this chapter is to develop the ideas of curvature and torsion, and this has been done by using the concept of the spin-vector of a vector.

The third chapter on surfaces deals first with the principal types of curves lying on them, and then goes on to discuss the concept of curvature for a surface. Particular reference is made to developable surfaces and ruled surfaces.

While the whole treatment is rigorous as far as is possible, in introducing the various ideas reference is constantly being made to the underlying geometrical intuition, and a large number of examples is included, together with their solutions, to help the reader to appreciate the ideas fully.

PREFACE

I cannot conclude without acknowledging the valuable suggestions and the interest shown by Dr. W. Ledermann while the manuscript was in the course of preparation. Any shortcomings must be attributed solely to myself. Finally I must thank the publishers, and in particular their printing staff, for the way in which they have carried out the translation of the manuscript into print.

K. L. WARDLE

Contents

Vector Results in Constant Use in the Text

1. The symbol $\mathbf{r}$ is used for a vector of length r. We write $r = |\mathbf{r}|$. For a unit vector in the direction of $\mathbf{r}$ we write $\hat{\mathbf{r}} \equiv \mathbf{r}/r$.

2. If $\mathbf{a},\mathbf{b}$ are vectors whose directions are inclined to one another at an angle θ (θ lying between 0 and π), we define the scalar product of $\mathbf{a},\mathbf{b}$ by $\mathbf{a}.\mathbf{b} = ab \cos \theta$; and if $\hat{\mathbf{c}}$ is a unit vector perpendicular to both $\mathbf{a}$ and $\mathbf{b}$ in such a sense that $\mathbf{a},\mathbf{b},\hat{\mathbf{c}}$ form a right-handed triad of vectors, we define the vector product of $\mathbf{a},\mathbf{b}$ by $\mathbf{a}\wedge\mathbf{b} = (ab \sin \theta)\hat{\mathbf{c}}$. If $\mathbf{b} = \mathbf{a}$ we have $\mathbf{a}.\mathbf{b} = \mathbf{a}.\mathbf{a} = a^2$ (and a common notation for $\mathbf{a}.\mathbf{a}$ is $\mathbf{a}^2$); also $\mathbf{a}\wedge\mathbf{b} = \mathbf{a}\wedge\mathbf{a} = \mathbf{0}$.

3. The triple scalar product $\mathbf{a}.(\mathbf{b}\wedge\mathbf{c})$, where the vectors $\mathbf{a},\mathbf{b},\mathbf{c}$ form a right-handed triad, is equal to six times the volume of the tetrahedron whose coterminous edges can be represented by the three vectors. We write

$$[\mathbf{a},\mathbf{b},\mathbf{c}] \equiv [\mathbf{abc}] = \mathbf{a}.(\mathbf{b}\wedge\mathbf{c}) = \mathbf{b}.(\mathbf{c}\wedge\mathbf{a}) = \mathbf{c}.(\mathbf{a}\wedge\mathbf{b}) = \\ -\,\mathbf{a}.(\mathbf{c}\wedge\mathbf{b}) = -\,\mathbf{b}.(\mathbf{a}\wedge\mathbf{c}) = -\,\mathbf{c}.(\mathbf{b}\wedge\mathbf{a}).$$

4. The triple vector product $\mathbf{a}\wedge(\mathbf{b}\wedge\mathbf{c})$ can be proved to have the expansion

$$\mathbf{a}\wedge(\mathbf{b}\wedge\mathbf{c}) = \mathbf{b}(\mathbf{a}.\mathbf{c}) - \mathbf{c}(\mathbf{a}.\mathbf{b}).$$

5. It can be shown that $[\mathbf{b}\wedge\mathbf{c},\mathbf{c}\wedge\mathbf{a},\mathbf{a}\wedge\mathbf{b}] = [\mathbf{a},\mathbf{b},\mathbf{c}]^2$. Provided that $[\mathbf{abc}] \neq 0$, and consequently

$$[\mathbf{b}\wedge\mathbf{c},\ \mathbf{c}\wedge\mathbf{a},\ \mathbf{a}\wedge\mathbf{b}] \neq 0,$$

we can resolve any vector $\mathbf{r}$ in the two ways:

$$\text{(i)}\quad \mathbf{r} = \frac{\mathbf{a}[\mathbf{rbc}] + \mathbf{b}[\mathbf{rca}] + \mathbf{c}[\mathbf{rab}]}{[\mathbf{abc}]};$$

$$\text{(ii)}\quad \mathbf{r} = \frac{(\mathbf{b}\wedge\mathbf{c})(\mathbf{r}.\mathbf{a}) + (\mathbf{c}\wedge\mathbf{a})(\mathbf{r}.\mathbf{b}) + (\mathbf{a}\wedge\mathbf{b})(\mathbf{r}.\mathbf{c})}{[\mathbf{abc}]}.$$

6. (i) $\mathbf{a}.\mathbf{b} = 0$ implies that either **a** is perpendicular to **b**, or that at least one of **a** and **b** is a zero vector.

(ii) $\mathbf{a} \wedge \mathbf{b} = \mathbf{0}$ implies that either **a** is parallel to **b**, or that at least one of them is zero.

(iii) $[\mathbf{abc}] = 0$ implies that **a**,**b**,**c** are all parallel to the same plane, or that at least one of them is zero.

(iv) $\mathbf{a} \wedge (\mathbf{b} \wedge \mathbf{c}) = \mathbf{0}$ implies that **a** is perpendicular to both **b** and **c**, or that **b** is parallel to **c**, or that one of the vectors is zero.

7. If **c** is a vector of constant length whose direction depends on a parameter t, so that **c** is a function of t, from the result that $\mathbf{c}^2 = \mathbf{c}.\mathbf{c}$, by differentiation with respect to t, we have $0 = \mathbf{c}.\dot{\mathbf{c}}$. so that $\dot{\mathbf{c}}$ is perpendicular to **c**.

CHAPTER ONE

Plane Curves

Introduction

In this first chapter we do not attempt to give anything in the nature of a complete account of plane curves, but we shall establish some of the ideas which reappear when we come to deal with curves in space. These ideas will be concerned with tangents, normals, curvature and envelopes. As the subsequent work will be dealt almost entirely by vector methods, these will be introduced in this chapter, though it must be remembered that vector work will only show to full advantage when applied to space of three dimensions.

Representation of a Plane Curve

In elementary work one first meets the equation of a curve in the form $y = f(x)$, an equation which, with the usual convention about functions, determines a single value of y for any given value of x. In dealing with conics one soon meets equations such as that of the parabola, $y^2 = 4ax$, where y is no longer a single-valued function, and a curve whose equation is of the form $f(x,y) = 0$ presents even further difficulties. We shall therefore consider curves whose equations are given in the parametric form $x = f(t), y = g(t)$, where t is a scalar parameter. Using $\mathbf{r}$ as the position vector from the origin to the point P on the curve whose parameter is t, we may then put $\mathbf{r} = \mathbf{i}x + \mathbf{j}y$, where $\mathbf{i}$ and $\mathbf{j}$ are unit vectors along the co-ordinate axes, and so $\mathbf{r} = \mathbf{r}(t)$, where $\mathbf{r}(t)$ is a vector function of t.

Arc-length

In elementary calculus the arc-length s between points whose parameters are t_1 and t_2 is given by

$$s = \int_{t_1}^{t_2} \left[\left(\frac{dx}{dt} \right)^2 + \left(\frac{dy}{dt} \right)^2 \right]^{1/2} dt.$$

This result evolved from an intuitive conception of arc-length in elementary work. But we find it convenient to regard such a formula as the definition of the arc-length. Also, to avoid confusion with notation to be used in a moment, we shall take u as the parameter, instead of t.

Now since we may write $d\mathbf{r}/du = \mathbf{i}dx/du + \mathbf{j}dy/du$ we get

$$\left(\frac{d\mathbf{r}}{du} \right)^2 = \frac{d\mathbf{r}}{du} \cdot \frac{d\mathbf{r}}{du} = \left(\frac{dx}{du} \right)^2 + \left(\frac{dy}{du} \right)^2,$$

and so we give

Definition 1.1. *The* **arc-length** *of a curve between points* P_1, P_2 *which correspond to the parameter values* u_1, u_2 *is defined by the relation*

$$s = \int_{u_1}^{u_2} \left[\frac{d\mathbf{r}}{du} \cdot \frac{d\mathbf{r}}{du} \right]^{1/2} du,$$

which is equivalent to saying

$$\frac{ds}{du} = \left[\frac{d\mathbf{r}}{du} \cdot \frac{d\mathbf{r}}{du} \right]^{1/2}.$$

Change of Parameter

The arc-length just defined does not depend on the choice of parameter, for if we take $u = \phi(v)$, where $\phi'(v)$ is non-zero in the range considered,

$$s = \int_{u_1}^{u_2}\left[\frac{d\mathbf{r}}{du}\cdot\frac{d\mathbf{r}}{du}\right]^{1/2} du = \int_{v_1}^{v_2}\left[\left(\frac{d\mathbf{r}}{dv}\cdot\frac{dv}{du}\right)\cdot\left(\frac{d\mathbf{r}}{dv}\cdot\frac{dv}{du}\right)\right]^{1/2}\frac{du}{dv}\,dv$$

$$= \int_{v_1}^{v_2}\left[\frac{d\mathbf{r}}{dv}\cdot\frac{d\mathbf{r}}{dv}\right]^{1/2} dv,$$

and in fact any intrinsic property of the curve, that is a property depending only on its shape and not upon its orientation nor upon the frame of reference, is not affected by the choice of the parameter in which it is expressed.

Tangent

We define the tangent to the curve at P to be the line in which lies the vector $d\mathbf{r}/du$, located at P.

Suppose $\mathbf{r}$ to be expressed in terms of the parameter s, the arc-length measured along the curve from some fixed point on it. In this case we may put $u = s$ in definition 1.1, from which we obtain at once

$$\frac{ds}{ds} = \left[\frac{d\mathbf{r}}{ds}\cdot\frac{d\mathbf{r}}{ds}\right]^{1/2} = 1,$$

so that

$$\frac{d\mathbf{r}}{ds}\cdot\frac{d\mathbf{r}}{ds} = 1.$$

It is now evident that $d\mathbf{r}/ds$ is a unit vector parallel to the tangent at P, and for this vector we shall always write $\mathbf{t}$. For this reason we shall normally use the arc-length s as the parameter, as this will lead to simplicity of expressions.

Normal

Since $\mathbf{t}.\mathbf{t} = 1$ we have $2\mathbf{t}.(d\mathbf{t}/ds) = 0$, so that $d\mathbf{t}/ds = d^2\mathbf{r}/ds^2$ is a vector perpendicular to $\mathbf{t}$. For the unit vector in the direction of $d^2\mathbf{r}/ds^2$ we shall write $\mathbf{p}$. The line through P, which is parallel to $\mathbf{p}$, we shall call the normal at P.

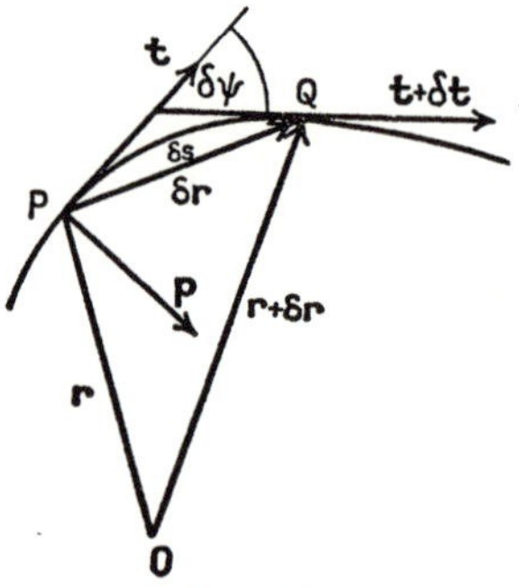

FIG. 1.1

Curvature

A natural way in which to measure the amount of bend in a curve is to take it to be the rate of turn of the tangent with respect to arc-length. So if the angle between 'neighbouring' tangents is $\delta\psi$ we shall call $\kappa = d\psi/ds$ the curvature of the curve at any point. For some positive scalar λ we have

$$\lambda\mathbf{p} = \frac{d^2\mathbf{r}}{ds^2} = \frac{d\mathbf{t}}{ds} = \frac{d\mathbf{t}}{d\psi}\frac{d\psi}{ds} = \kappa\frac{d\mathbf{t}}{d\psi}.$$

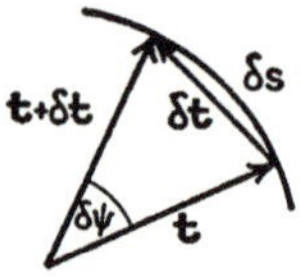

FIG. 1.2

But as $\mathbf{t}$ is a unit vector its 'neighbouring' vector $\mathbf{t} + \delta\mathbf{t}$ is also a unit vector, and we see at once, from figure 1.2, that, to a first approximation $|\delta\mathbf{t}| = 1.\delta\psi$, so that, taking limits, $d\mathbf{t}/d\psi$ is a unit vector perpendicular to $\mathbf{t}$.

Hence $\lambda = \pm\kappa$, and $\mathbf{p} = \pm d\mathbf{t}/d\psi$, the positive or negative signs being taken according as $d\psi/ds$ is positive or negative.

Since

$$|\kappa\mathbf{p}| = \left|\frac{d^2\mathbf{r}}{ds^2}\right| = \left|\mathbf{i}\frac{d^2x}{ds^2} + \mathbf{j}\frac{d^2y}{ds^2}\right|$$

we have at once

$$\kappa^2 = \left(\frac{d^2x}{ds^2}\right)^2 + \left(\frac{d^2y}{ds^2}\right)^2,$$

and from this we can obtain the usual expression for the curvature of a curve $y = f(x)$ by taking x as the parameter instead of s.

Using $y_1 \equiv dy/dx$, and $y_2 \equiv d^2y/dx^2$, from the relation $ds/dx = (1 + y_1^2)^{1/2}$ we obtain $dx/ds = (1 + y_1^2)^{-1/2}$, so that

$$\frac{d^2x}{ds^2} = \frac{dx}{ds}\frac{d}{dx}\left(\frac{dx}{ds}\right) = (1 + y_1^2)^{-1/2}[-\tfrac{1}{2}(1 + y_1^2)^{-3/2}2y_1y_2]$$

$$= -y_1y_2(1 + y_1^2)^{-2},$$

$$\frac{d^2y}{ds^2} = \frac{dx}{ds}\frac{d}{dx}\left(\frac{dy}{dx}\frac{dx}{ds}\right)$$

$$= (1 + y_1^2)^{-1/2}\frac{d}{dx}[y_1(1 + y_1^2)^{-1/2}]$$

$$= (1 + y_1^2)^{-1/2}[y_2(1 + y_1^2)^{-1/2} - y_1^2y_2(1 + y_1^2)^{-3/2}]$$

$$= y_2(1 + y_1^2)^{-2}[(1 + y_1^2) - y_1^2].$$

Thus

$$\kappa^2 = y_2^2(1 + y_1^2)^{-4}(1 + y_1^2),$$

and so

$$\kappa^2 = \frac{(d^2y/dx^2)^2}{[1 + (dy/dx)^2]^3}.$$

Since for a circle of radius ρ we may write $\delta s = \rho\delta\psi$, we have in that case

$$\kappa = d\psi/ds = 1/\rho.$$

Circle of Curvature

Since a unique circle can be drawn through any three non-collinear points, another way of measuring the rate of bend of a plane curve would appear to be to take it as the rate of bend, or curvature, of the limiting position of such a circle as the three points move into coincidence, if such a limit can be found. There are, however, difficulties here which we shall not discuss. Instead we talk of two curves, whose position vectors are $\mathbf{r}_1(u)$ and $\mathbf{r}_2(v)$, having three-point contact at some point if there are values u_0,v_0 of the parameters u,v for which

$$\mathbf{r}_1(u_0) = \mathbf{r}_2(v_0), \dot{\mathbf{r}}_1(u_0) = \dot{\mathbf{r}}_2(v_0), \ddot{\mathbf{r}}_1(u_0) = \ddot{\mathbf{r}}_2(v_0),$$

where dots denote differentiation with respect to the parameters.

The equation of a circle of radius ρ, whose centre C is at the extremity of the vector $\mathbf{c}$, may be written

$$[\mathbf{r}_2(u) - \mathbf{c}] \cdot [\mathbf{r}_2(u) - \mathbf{c}] = \rho^2$$

where u is some parameter on which $\mathbf{r}$ depends. For this circle to have three-point contact at some point P with a curve the points of which lie at the extremities of the position vector $\mathbf{r}_1(s)$, we must have, for values u_0,s_0 of the parameters at P,

$$\mathbf{r}_1(s_0) = \mathbf{r}_2(u_0), \dot{\mathbf{r}}_1(s_0) = \dot{\mathbf{r}}_2(u_0), \ddot{\mathbf{r}}_1(s_0) = \ddot{\mathbf{r}}_2(u_0).$$

Now, from the equation of the circle, we obtain by differentiation

$$2\dot{\mathbf{r}}_2(u) \cdot [\mathbf{r}_2(u) - \mathbf{c}] = 0,$$
$$2\ddot{\mathbf{r}}_2(u) \cdot [\mathbf{r}_2(u) - \mathbf{c}] + 2\dot{\mathbf{r}}_2(u) \cdot \dot{\mathbf{r}}_2(u) = 0,$$

and using the first of these equations with the relations above we see that $\dot{\mathbf{r}}_1(s_0)$ is perpendicular to $[\mathbf{r}_2(u_0) - \mathbf{c}]$, showing that the centre of the circle lies on the normal to the curve at P. From the second equation, since $\dot{\mathbf{r}}_2(u_0) = \dot{\mathbf{r}}_1(s_0)$ is $\mathbf{t}$, the unit tangent vector at P to the curve, the term $\dot{\mathbf{r}}_2(u_0).\dot{\mathbf{r}}_2(u_0)$ is unity. Also $\ddot{\mathbf{r}}_2(u_0) = \ddot{\mathbf{r}}_1(s_0) = \pm \kappa\mathbf{p}$, where $\mathbf{p}$ is the unit normal vector at P, and κ is the curvature of the curve. But $[\mathbf{r}_2(u_0) - \mathbf{c}]$ is the vector $\overrightarrow{CP}$, so we have $|\kappa|\rho = 1$, or $|\kappa| = 1/\rho$. So we give

Definition 1.2. *The circle having three-point contact with a curve at a given point P on the curve is called the* **circle of curvature** *at P. Its centre is called the* **centre of curvature,** *and the reciprocal of its radius gives the measure of the curvature of the curve at P.*

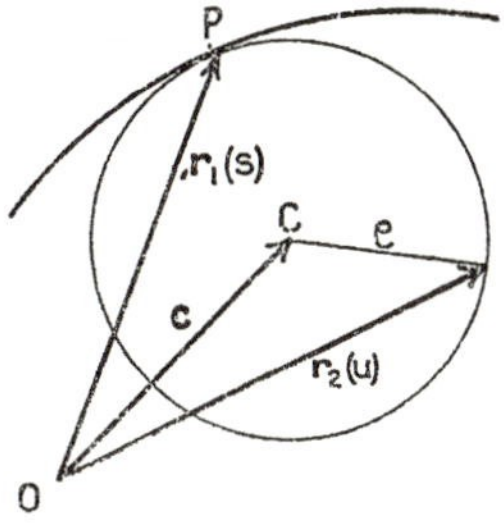

FIG. 1.3

The Envelope of a Family of Straight Lines

In drawing the curve represented by the equation $y = f(x)$ one plots a series of points corresponding to the pairs of values of x and y, and then attempts to join these points by a smooth curve. If the curve could be drawn mechanically by the motion of a point, say by regarding time as the parameter when the curve is expressed in parametric form, we should expect our previous drawing to approximate more closely to this curve as we increased the number of points on our drawing.

Now the tangents to a curve form a one-parameter family of straight lines whose cartesian equations are of the form

$$a(u)\,x + b(u)\,y + c(u)\,z = 0,$$

u being the parameter.

Just as we attempt to draw a smooth curve passing through a series of points, we may attempt to draw a smooth curve touching the series of tangents given by taking a succession of values of the parameter u.

Now the limit of the point of intersection of two 'neighbouring' tangents will satisfy both the equation above and also the equation

$$\dot{a}(u)\,x + \dot{b}(u)\,y + \dot{c}(u)\,z = 0.†$$

The locus of these limiting points is found by eliminating u between the two equations above, if possible, and will be the curve itself, and the curve is now regarded as being enveloped by its tangents.

As an illustration, consider the family of tangents to the parabola $y^2 = 4ax$. The equation of the tangent at the point $(au^2,2au)$ on the curve is $yu = x + au^2$, and, differentiating with respect to u we get $y = 2au$. Eliminating u gives $y^2/2a = x + y^2/4a$, which reduces to $y^2 = 4ax$, the equation of the parabola itself.

Definition 1.3. *The* **envelope** *of a one-parameter family of straight lines is the set of points satisfying both the equations* $(x,y,u) = 0$ *and* $\partial f(x,y,u)/\partial u = 0$, *where the first equation is that of the lines of the family.*

Generally, for any curve, we may write the equation of an arbitrary tangent in the form $\mathbf{r.n} = p$, where $\mathbf{r}$ is the position

† For if $L(x,y,u) = 0$, $L(x,y,u + \delta u) = 0$ are 'neighbouring' lines belonging to the family then $[L(x,y,u + \delta u) - L(x,y,u)]/\delta u = 0$ is the equation of some line through their point of intersection; so $\partial L(x,y,u)/\partial u = 0$ is the equation of a line through the limiting position of this point.

vector of a point on the tangent, p is the length of the perpendicular from the origin on to the tangent, and $\mathbf{n}$ is a unit vector drawn from the origin along this perpendicular. Of course $\mathbf{n}$ and p are functions of a parameter u.

By definition the envelope of this tangent is the set of points satisfying both $\mathbf{r}.\mathbf{n} = p$, and $\mathbf{r}.\dot{\mathbf{n}} = \dot{p}$.

Now if ψ is the angle made by $\mathbf{n}$ with some fixed direction, and if $\mathbf{m}$ is a unit vector perpendicular to $\mathbf{n}$ in the appropriate sense, then, as $d\mathbf{n}/d\psi$ is a unit vector perpendicular to $\mathbf{n}$, we have $d\mathbf{n}/d\psi = (d\mathbf{n}/du)(du/d\psi) = \mathbf{m}$, so that $\dot{\mathbf{n}} = \mathbf{m}\dot{\psi}$, and so from the second equation we get $\mathbf{r}.\mathbf{m}\dot{\psi} = \dot{p}$. But since any vector in the plane can be expressed as a linear combination of two independent vectors we may write $\mathbf{r}$ in the form $\lambda\mathbf{n} + \mu\mathbf{m}$, where λ,μ are scalars. Then we have

$$p = \mathbf{r}.\mathbf{n} = \lambda\, \mathbf{n}.\mathbf{n} + \mu\, \mathbf{m}.\mathbf{n} = \lambda,$$

$$dp/d\psi = \mathbf{r}.\mathbf{m} = \lambda\, \mathbf{n}.\mathbf{m} + \mu\, \mathbf{m}.\mathbf{m} = \mu,$$

since $\mathbf{n}.\mathbf{m} = 0$, and so the equation of the envelope is

$$\mathbf{r} = \mathbf{n}\, p + \mathbf{m}\, dp/d\psi.$$

Notice that, by Fig. 1.4, $\mu = (dp/d\psi)$ is the length of the perpendicular from the origin to the normal to the curve at the point of contact of the tangent.

If we repeat this process, starting with the normals to the curve, then the equation of the envelope of the normals is given by the values of $\mathbf{r}$ satisfying $\mathbf{r}.\mathbf{m} = dp/d\psi$, $\mathbf{r}.\dot{\mathbf{m}} = \dot{\overline{dp/d\psi}}$. This last equation may be written $-\mathbf{r}.(\mathbf{n}\,\dot{\psi}) = (d^2p/d\psi^2)\,\dot{\psi}$, since we can show that $\dot{\mathbf{m}} = -\mathbf{n}\dot{\psi}$ in the same way that we showed that $\dot{\mathbf{n}} = \mathbf{m}\dot{\psi}$, and proceeding as before, from $\mathbf{r} = \nu\mathbf{n} + \pi\mathbf{m}$, where ν,π are scalars, we get

$$dp/d\psi = \mathbf{r}.\mathbf{m} = \nu\, \mathbf{n}.\mathbf{m} + \pi\, \mathbf{m}.\mathbf{m} = \pi,$$

$$-\, d^2p/d\psi^2 = \mathbf{r}.\mathbf{n} = \nu\, \mathbf{n}.\mathbf{n} + \pi\, \mathbf{m}.\mathbf{n} = \nu.$$

So the equation of the envelope of the normals to a curve, a locus known as the **evolute**, is

$$\mathbf{r} = \mathbf{m}\,(dp/d\psi) - \mathbf{n}\,(d^2p/d\psi^2),$$

and, from what has been said earlier about the circle of curvature, it is now evident that the evolute is the locus of the centres of curvature, and that the radius of curvature is given by

$$\rho = p + d^2p/d\psi^2.$$

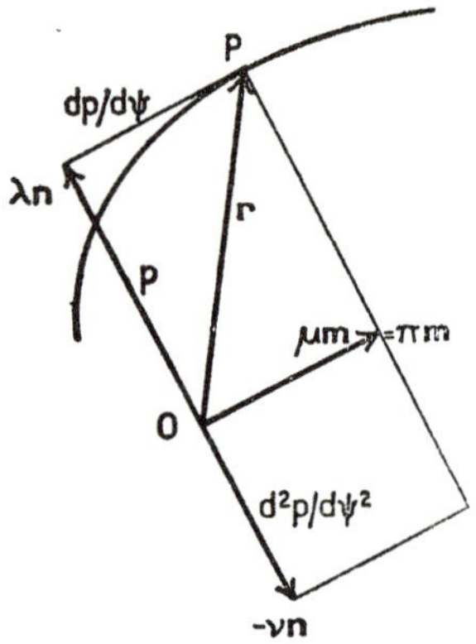

FIG. 1.4

EXAMPLES ON CHAPTER ONE

1.1. Prove that

$$|\kappa| = |\mathbf{t}\wedge(d\mathbf{t}/ds)| = \left|\frac{d^2y}{ds^2}\frac{dx}{ds} - \frac{d^2x}{ds^2}\frac{dy}{ds}\right|.$$

1.2. Deduce the parametric formula for curvature from

$$|\kappa| = |\mathbf{t}\wedge(\dot{\mathbf{t}}/\dot{s})|.$$

1.3. If, for some curve Γ, s is arc-length, r the length of the position vector $\overrightarrow{OP}$ of a point P on Γ, and ϕ the angle between OP and the tangent $\mathbf{t}$ at P, show that $\cos\phi = dr/ds$.

1.4. If p is the length of the perpendicular from the origin on to the tangent at P on a curve Γ, and if $\mathbf{n}$ is the unit normal at P, show that $p = |\mathbf{r}.\mathbf{n}|$, and by differentiation with respect to s prove that

$$|\kappa| = \left|\frac{1}{r}\frac{dp}{dr}\right|,$$

where r is the length of $\mathbf{r}$.

1.5. Prove that

$$r^2 = p^2 + \left(\frac{dp}{d\psi}\right)^2.$$

1.6. If the foot of the perpendicular p from the origin O on to a tangent to a curve Γ lies always on a fixed line at distance a from O, and if ϕ is the angle between p and the perpendicular from O to the fixed line, use the result of example 1.5 to show that

$$1 + \cos 2\phi = 2a/r.$$

Deduce that the angle between p and the radius vector is also ϕ.

Taking cartesian axes along the fixed line and the perpendicular to it through O, show that the equation of Γ may be written $y^2 = 4ax$.

CHAPTER TWO

Curves in Space

Introduction

In dealing with a curve in space of two dimensions the advantages of being able to express the equation of the curve $f(x,y) = 0$ in a parametric form $x = x(t)$, $y = y(t)$ are quickly realized. When we come to curves lying in space of three dimensions one method of description would be to regard them as the intersections of two surfaces. This would require two equations $f(x,y,z) = 0$, $g(x,y,z) = 0$ for their specification, and the advantages of a parametric form $x = x(t)$, $y = y(t)$, $z = z(t)$ are even more apparent, especially as the idea extends for curves in space of any number of dimensions.

Physically the parametric form is suggested at once by considering the curve as being the path of a particle moving in space whose position (x,y,z) is given in terms of the time t.

The problem of finding such a parametric form from a pair of equations is by no means straightforward, and we shall restrict our work to curves whose points can be specified by position vectors $\mathbf{r} = \mathbf{i}x + \mathbf{j}y + \mathbf{k}z$, where x, y, z, and so $\mathbf{r}$, are functions of a parameter t, and where $\mathbf{i}$, $\mathbf{j}$, $\mathbf{k}$ are three mutually orthogonal fixed unit vectors at the origin, independent, of course, of t. We give therefore

Definition 2.1. *A* **curve** *Γ in Euclidean space of three dimensions is the locus of the point P at the extremity of a position vector* $\mathbf{r}$, *where* $\mathbf{r}$ *is a function of the single parameter t.*

We shall assume that $\mathbf{r} = \mathbf{r}(t)$ satisfies the conditions

(a) $\mathbf{r}(t)$ is single-valued,

(b) $d\mathbf{r}/dt$ exists and is not zero,

(c) all the successive derivatives of $\mathbf{r}$ exist and are continuous, except possibly at a finite number of special points.

The parameter t may take all values, or be limited to some particular range of values.

Since $\mathbf{r} = \mathbf{i}x + \mathbf{j}y + \mathbf{k}z$, the cartesian co-ordinates x, y, z of P on Γ will all be functions of t, and we shall assume that, for the curves with which we deal, we can write down a Taylor expansion for $\mathbf{r}(t)$

$$\mathbf{r}(t) = \mathbf{r}(t_0) + (t - t_0)\dot{\mathbf{r}}(t_0) + \\ + (t - t_0)^2\ddot{\mathbf{r}}(t_0)/2! + (t - t_0)^3\dddot{\mathbf{r}}(t_0)/3! + \ldots,$$

where the dots denote derivatives of $\mathbf{r}$ with respect to t, and the suffixes indicate their values at some point where $t = t_0$.

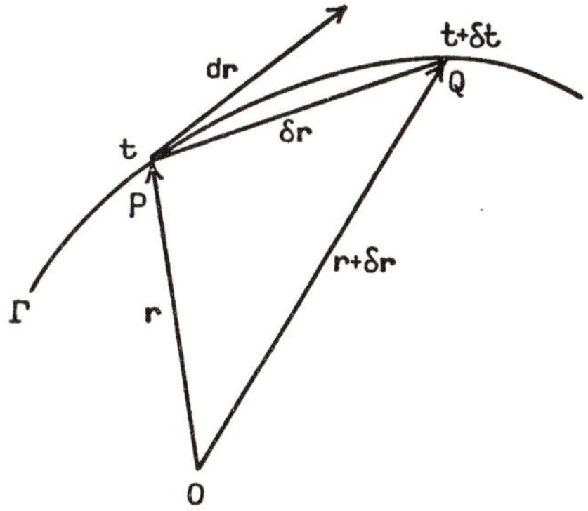

FIG. 2.1

Differences and Differentials

Suppose P, Q are the points on Γ whose parameters are t and $t + \delta t$, and whose position vectors are $\mathbf{r}$ and $\mathbf{r} + \delta\mathbf{r}$. Then $\overrightarrow{PQ}$ is the vector

$$(\mathbf{r} + \delta\mathbf{r}) - \mathbf{r} = \delta\mathbf{r} = \mathbf{r}(t + \delta t) - \mathbf{r}(t).$$

Now

$$\mathbf{r}(t + \delta t) = \mathbf{r}(t) + \delta t\, \dot{\mathbf{r}}(t) + \delta t^2\, \ddot{\mathbf{r}}(t)/2! + \ldots,$$

and so

$$\delta\mathbf{r} = \dot{\mathbf{r}}\, \delta t + \ddot{\mathbf{r}}\, \delta t^2/2! + \ldots.$$

Definition 2.2. $\delta\mathbf{r}$ *is called the* **difference** *of* $\mathbf{r}$ *at* t.

A first approximation to this difference when δt is small is $\dot{\mathbf{r}}\, \delta t$, and we recall that the differential of a scalar function $f(t)$ is $df = (df/dt)\, \delta t$, so we give

Definition 2.3. $\dot{\mathbf{r}}\, \delta t$ *is called the* **differential** *of* $\mathbf{r}$ *at* t *and is denoted by* $d\mathbf{r}$.

As t is an independent variable its own differential is given by $dt = (dt/dt)\, \delta t = \delta t$, so it is immaterial whether we write δt or dt. So $d\mathbf{r} = (d\mathbf{r}/dt)\, dt = \dot{\mathbf{r}}\, dt$.

It is important to point out that $\delta\mathbf{r}$ and $d\mathbf{r}$ depend on both δt and t, and also that there is nothing in the definition to imply that δt is necessarily small, so the vector $d\mathbf{r}$ may have any magnitude we like.

Tangent

From geometrical considerations the direction of PQ as Q approaches P becomes closer to the direction of the tangent at P. We therefore give

Definition 2.4. *A non-zero vector located at* P *on* Γ *and having the direction of* $d\mathbf{r}/dt$ *is called a* **tangential vector** *to* Γ *at* P.

Thus $d\mathbf{r}$ located at P is a tangential vector.

The condition (b) that $\dot{\mathbf{r}} \neq \mathbf{0}$ implies that these tangential vectors exist.

Length of Arc

Suppose AB is an arc of a curve Γ. By taking a succession of points on the arc as vertices, a polygon can be inscribed to the arc with A and B as extreme vertices. As the number of vertices is increased it is intuitive that the perimeter of this polygon will have a length approximating to the length of the arc.

Now the length of a typical side PQ of such a polygon is $|\delta\mathbf{r}|$, so the length of the whole perimeter will be $\Sigma(\delta\mathbf{r}.\delta\mathbf{r})^{1/2}$. From the theory of integration, under suitable conditions to be found in texts on Analysis, the lengths of the perimeters of all inscribed polygons will have an upper bound, and, as the lengths of the sides PQ approach zero, the sum $\Sigma(\delta\mathbf{r}.\delta\mathbf{r})^{1/2}$ will approach

$$\int (\dot{\mathbf{r}}.\dot{\mathbf{r}})^{1/2}\, dt.$$

We therefore give

Definition 2.5. *The* **arc-length** *of a curve Γ between the points A and B is given by*

$$s = \int_{t_0}^{t} (\dot{\mathbf{r}}.\dot{\mathbf{r}})^{1/2}\, dt = \int_{t_0}^{t} (\dot{x}^2 + \dot{y}^2 + \dot{z}^2)^{1/2}\, dt,$$

where t_0, t are the parameter values of the points A and B.

Element of Arc

With this definition we have $ds/dt = (\dot{\mathbf{r}}.\dot{\mathbf{r}})^{1/2}$, and so we give

Definition 2.6. *The differential $ds = (\dot{\mathbf{r}}.\dot{\mathbf{r}})^{1/2}\, dt$ is called the* **element of arc.**

Thus $ds = (d\mathbf{r}.d\mathbf{r})^{1/2}$; and so ds is the length of the tangential vector $d\mathbf{r}$. Hence $d\mathbf{r}/ds$ is a unit tangent vector, called simply the tangent vector. We shall use dashes to denote differentiation with respect to s, the arc-length, and so write $\mathbf{r}' = d\mathbf{r}/ds$.

Theorem 2.1. *As PQ decreases to zero length the ratio (chord PQ/arc PQ) approaches unity.* For

$$\lim_{PQ\to 0}\frac{\text{chord } PQ}{\text{arc } PQ} = \lim_{\delta t\to 0}\frac{(\delta\mathbf{r}.\delta\mathbf{r})^{1/2}}{\delta s}$$

$$= \lim_{\delta t\to 0}\left[\frac{\delta\mathbf{r}.\delta\mathbf{r}}{\delta t.\delta t}\right]^{1/2}\frac{\delta t}{\delta s}$$

$$= (\dot{\mathbf{r}}.\dot{\mathbf{r}})^{1/2}\frac{dt}{ds} = 1.$$

Theorem 2.2. *The length of arc is independent of the parameter chosen.*

Suppose $t = \phi(u)$, where $\phi(u)$ is a single-valued function with a continuous positive derivative. Then

$$s = \int_{t_0}^{t}(\dot{\mathbf{r}}.\dot{\mathbf{r}})^{1/2}dt = \int_{t_0}^{t}\left[\frac{d\mathbf{r}}{du}\cdot\frac{d\mathbf{r}}{du}\right]^{1/2}\frac{du}{dt}\,dt$$

$$= \int_{u_0}^{u}\left[\frac{d\mathbf{r}}{du}\cdot\frac{d\mathbf{r}}{du}\right]^{1/2}\frac{du}{dt}\,\phi'(u)\,du$$

$$= \int_{u_0}^{u}\left[\frac{d\mathbf{r}}{du}\cdot\frac{d\mathbf{r}}{du}\right]^{1/2}du,$$

since $\phi'(u) = dt/du$.

Spin-vector of a Vector

Suppose the vector **v** to be a function of some parameter t. Then, as t varies, **v** will alter both in direction and magnitude, and so will possess a rate of turn; and if **v** is regarded as located at some origin, for all values of t, it will also possess an instantaneous axis of rotation for any particular value of t. We shall therefore seek a vector whose magnitude measures the rate of turn, and whose direction is that of the axis of rotation just mentioned.

Let $\mathbf{v}, \mathbf{v} + \delta\mathbf{v}$ be the vectors corresponding to the values $t, t + \delta t$ of the parameter. Then $\mathbf{v} \wedge (\mathbf{v} + \delta\mathbf{v})$ is a vector, perpendicular to both, of magnitude $|\mathbf{v}||\mathbf{v} + \delta\mathbf{v}| \sin \delta\theta$, where

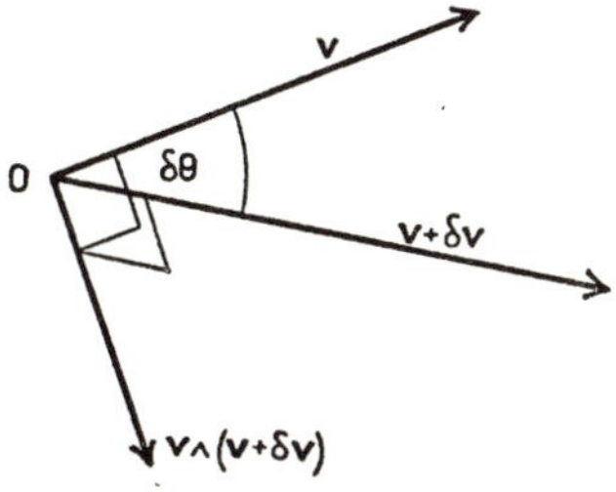

FIG. 2.2

$\delta\theta$ is the angle between the two vectors. So, as δt approaches zero, the vector

$$\lim_{\delta t \to 0} \frac{\mathbf{v} \wedge (\mathbf{v} + \delta\mathbf{v})}{|\mathbf{v}||\mathbf{v} + \delta\mathbf{v}|} \frac{1}{\delta t}$$

is a vector parallel to the axis of turn, whose magnitude is

$$\lim_{\delta t \to 0} \frac{|\mathbf{v}||\mathbf{v} + \delta\mathbf{v}|}{|\mathbf{v}||\mathbf{v} + \delta\mathbf{v}|} \frac{\sin \delta\theta}{\delta\theta} \frac{\delta\theta}{\delta t} = \frac{d\theta}{dt} = \dot{\theta},$$

the rate of turn of $\mathbf{v}$ with respect to t, and this result is clearly independent of the magnitude of $\mathbf{v}$. Now the first limit is

$$\lim_{\delta t \to 0} \frac{\mathbf{v} \wedge \delta\mathbf{v}}{|\mathbf{v}||\mathbf{v} + \delta\mathbf{v}|} \frac{1}{\delta t} = \frac{\mathbf{v} \wedge \dot{\mathbf{v}}}{\mathbf{v}^2},$$

$$(\mathbf{v}^2 \equiv \mathbf{v}.\mathbf{v} = v^2 \text{ and } \dot{\mathbf{v}} = d\mathbf{v}/dt)$$

so we have

Definition 2.7. *The* **spin-vector** *of* $\mathbf{v}(t)$ *is defined as* $(\mathbf{v} \wedge \dot{\mathbf{v}})/\mathbf{v}^2$, *where* $\dot{\mathbf{v}} = d\mathbf{v}/dt$.

Curvature of a Curve in Space

Using the idea just developed, and taking s the arc-length as parameter, the spin-vector of the unit tangent $\mathbf{t}$ at a point P on a curve Γ is the vector $(\mathbf{t} \wedge \mathbf{t}')/\mathbf{t}^2 = (\mathbf{t} \wedge \mathbf{t}')$, where $\mathbf{t}' = d\mathbf{t}/ds$. We therefore give

Definition 2·8. *The rate of turn of the tangent with respect to arc at a point P of a curve Γ is called the* **curvature** κ. *So*

$$\kappa = |\mathbf{t} \wedge \mathbf{t}'| = |\mathbf{r}' \wedge \mathbf{r}''| .$$

Note that κ is essentially positive.

Since $\mathbf{t}$ is turning instantaneously about the vector $\mathbf{t} \wedge \mathbf{t}'$ it is doing so in a plane perpendicular to this vector. The two normals at P to Γ, one in this plane, one along the vector $\mathbf{t} \wedge \mathbf{t}'$, are prominent in the further theory, and so we give

Definition 2.9. *The unit normal vector* $\mathbf{p}$ *to Γ at P in the plane of turn of* $\mathbf{t}$ *and drawn from P in the direction of motion of the extremity of the localized vector* $\mathbf{t}$ *is called the* **principal normal** *at P*, and also

Definition 2.10. *The unit normal vector* $\mathbf{b}$ *drawn at P on Γ in the direction of the vector* $(\mathbf{t} \wedge \mathbf{t}')$ *is called the* **binormal** *at P.*

The unit vectors $\mathbf{t}, \mathbf{p}, \mathbf{b}$ then form a right-handed triad of perpendicular vectors, and from the definition of κ we get $\kappa\mathbf{b} = \mathbf{t} \wedge \mathbf{t}' = \mathbf{r}' \wedge \mathbf{r}''$, and
$\kappa\mathbf{p} = \kappa\mathbf{b} \wedge \mathbf{t} = (\mathbf{r}' \wedge \mathbf{r}'') \wedge \mathbf{r}' = \mathbf{r}''(\mathbf{r}')^2 - \mathbf{r}'(\mathbf{r}''.\mathbf{r}') = \mathbf{r}'' = \mathbf{t}'$, since $|\mathbf{r}'| = 1$, and so $\mathbf{r}''.\mathbf{r}' = 0$.

Torsion of a Curve in Space

From our definition, the spin-vector of $\mathbf{b}$ is $\mathbf{b} \wedge \mathbf{b}'$, but as yet we know nothing of the direction of this vector, for although $\mathbf{b}'$, the derivative of the unit vector $\mathbf{b}$, is known to be perpendicular to $\mathbf{b}$, its actual direction is not yet known. We therefore calculate the spin-vector of $\mathbf{b}$ by using the parallel vector $\mathbf{r}' \wedge \mathbf{r}''$, $\mathbf{r}'$ and $\mathbf{r}''$ being respectively parallel to $\mathbf{t}$ and $\mathbf{p}$. This gives

$$\frac{(\mathbf{r}' \wedge \mathbf{r}'') \wedge (\mathbf{r}' \wedge \mathbf{r}'')'}{|\mathbf{r}' \wedge \mathbf{r}''|^2} = \frac{(\mathbf{r}' \wedge \mathbf{r}'') \wedge (\mathbf{r}' \wedge \mathbf{r}''')}{\kappa^2}$$

$$(\text{since } \mathbf{r}'' \wedge \mathbf{r}'' = 0)$$

$$= \frac{\mathbf{r}'[\mathbf{r}',\mathbf{r}'',\mathbf{r}'''] - \mathbf{r}'''[\mathbf{r}',\mathbf{r}'',\mathbf{r}']}{\kappa^2}$$

$$= \mathbf{t}[\mathbf{r}',\mathbf{r}'',\mathbf{r}''']/\kappa^2,$$

the second triple scalar product being zero since it contains two vectors alike. So the binormal turns about an axis along the tangent at P, and we give

Definition 2.11. *The rate of turn of the binormal with respect to arc at a point P of a curve Γ is called the* **torsion** λ.

The torsion λ measures a second rate of bending for the curve, and we get

$$\lambda = [\mathbf{r}',\mathbf{r}'',\mathbf{r}''']/\kappa^2 = [\mathbf{t},\mathbf{t}',\mathbf{t}'']/\kappa^2.$$

Now the sign of λ depends on the sign of the triple scalar product, so whereas κ, the curvature, is essentially positive, λ, the torsion, may take either sign.

We have also found the direction of $\mathbf{b} \wedge \mathbf{b}'$, for this vector is $\lambda\mathbf{t}$.

Spin-vector of p

Since $\mathbf{p} = \mathbf{b} \wedge \mathbf{t}$, its spin-vector $\mathbf{p} \wedge \mathbf{p}'$ is

$$(\mathbf{b} \wedge \mathbf{t}) \wedge (\mathbf{b} \wedge \mathbf{t})' = (\mathbf{b} \wedge \mathbf{t}) \wedge (\mathbf{b} \wedge \mathbf{t}' + \mathbf{b}' \wedge \mathbf{t})$$

$$= \mathbf{b}[\mathbf{b},\mathbf{t},\kappa\mathbf{p}] - \mathbf{t}'[\mathbf{b},\mathbf{t},\mathbf{b}] +$$

$$+ \mathbf{b}'[\mathbf{b},\mathbf{t},\mathbf{t}] - \mathbf{t}[\mathbf{b},\mathbf{t},\mathbf{b}'].$$

Now the second and third of the triple scalar products are zero, having two vectors alike in each, and the fourth is $-\mathbf{t}.(\mathbf{b} \wedge \mathbf{b}')$, which is $-\mathbf{t}.\lambda\mathbf{t}$, that is $-\lambda$. So, as $[\mathbf{t},\mathbf{p},\mathbf{b}] = 1$,

$$\mathbf{p} \wedge \mathbf{p}' = \lambda\mathbf{t} + \kappa\mathbf{b}.$$

Now the spin-vector of $\mathbf{b}$ is along $\mathbf{t}$, the spin-vector of $\mathbf{t}$ is along $\mathbf{b}$ and the spin-vector of $\mathbf{p}$ is along $\lambda\mathbf{t} + \kappa\mathbf{b}$, a vector which is known as the Darboux vector. So the whole triad of vectors, $\mathbf{t}$, $\mathbf{p}$, $\mathbf{b}$ has instantaneous components of spin $(\lambda, 0, \kappa)$ about $\mathbf{t}$, $\mathbf{p}$, $\mathbf{b}$ respectively, and these measure the rates of turn per arc about these vectors.

Twisted curves in space are therefore sometimes referred to as curves with double curvature, there being instantaneous spin of the $\mathbf{t}$, $\mathbf{p}$, $\mathbf{b}$, triad about $\mathbf{t}$ and $\mathbf{b}$, but not about $\mathbf{p}$. These results show up more clearly when we discuss the form of the curve at any point in relation to the triad of vectors, $\mathbf{t}$, $\mathbf{p}$, $\mathbf{b}$.

The Frenet–Serret Formulae

To investigate the properties of curves in space we shall need the derivatives of the vectors $\mathbf{t}$, $\mathbf{p}$, $\mathbf{b}$. We have already seen that

$$\mathbf{t}' = \kappa\mathbf{p}.$$

From $\mathbf{p} \wedge \mathbf{p}' = \lambda\mathbf{t} + \kappa\mathbf{b}$, taking vector products with $\mathbf{p}$, we get

$$(\mathbf{p} \wedge \mathbf{p}') \wedge \mathbf{p} = \mathbf{p}'(\mathbf{p}.\mathbf{p}) - \mathbf{p}(\mathbf{p}'.\mathbf{p}) = \mathbf{p}',$$

for $\mathbf{p}.\mathbf{p} = 1$, and so $\mathbf{p}.\mathbf{p}' = 0$. This gives

$$\mathbf{p}' = (\lambda\mathbf{t} + \kappa\mathbf{b}) \wedge \mathbf{p} = \lambda\mathbf{b} - \kappa\mathbf{t} = -\kappa\mathbf{t} + \lambda\mathbf{b}.$$

Similarly, from $\mathbf{b} \wedge \mathbf{b}' = \lambda\mathbf{t}$, taking vector products with $\mathbf{b}$, we get

$$(\mathbf{b} \wedge \mathbf{b}') \wedge \mathbf{b} = \mathbf{b}'(\mathbf{b}.\mathbf{b}) - \mathbf{b}(\mathbf{b}'.\mathbf{b}) = \mathbf{b}',$$

for $\mathbf{b}.\mathbf{b} = 1$, and so $\mathbf{b}.\mathbf{b}' = 0$. This gives

$$\mathbf{b}' = \lambda\mathbf{t} \wedge \mathbf{b} = -\lambda\mathbf{p}.$$

So we have the Frenet–Serret

Theorem 2.3.

$$\begin{aligned} \mathbf{t}' &= \kappa\mathbf{p}, \\ \mathbf{p}' &= -\kappa\mathbf{t} + \lambda\mathbf{b}, \\ \mathbf{b}' &= -\lambda\mathbf{p}. \end{aligned}$$

Of course, for a plane curve, **b** is constant, and so $\mathbf{b}' = \mathbf{0}$, that is $\lambda = 0$.

Geometrical Properties

Definition 2.12. *The three planes formed by the pairs of vectors* **t**,**p**; **p**,**b**; **b**,**t**; *at a point* P *of a curve* Γ *are called respectively the* **osculating plane**, *the* **normal plane**, *and the* **rectifying plane.**

It is at once evident that all lines through P in the normal plane are perpendicular to **t**, and so are called normals to the curve at P, and these lie in the normal plane.

It would appear that any plane through P cuts Γ in at least one point, P; that any plane through the tangent vector **t** would seem to have at least two coincident intersections with Γ at P; and that to get even closer contact at least three such coincident intersections with Γ at P would be required. We shall prove rigorously

Theorem 2.4. *The plane through* P *having three-point contact with* Γ *at* P *is the osculating plane.*

If **R** is the position vector of any point on the osculating plane, and if **r** is the position vector of P, then $(\mathbf{R} - \mathbf{r})$ lies in the plane of **t** and **p**. So the equation of the osculating plane is $[(\mathbf{R} - \mathbf{r}), \mathbf{t}, \mathbf{p}] = 0$, and since $\mathbf{t} = \mathbf{r}'$, $\kappa\mathbf{p} = \mathbf{r}''$, this is equivalent to

$$[(\mathbf{R} - \mathbf{r}), \mathbf{r}', \mathbf{r}''] = 0.$$

We now show that for a plane to have three-point contact with Γ at P its equation must be the one just given.

If $\mathbf{n}$ is the perpendicular from the origin on to a plane, the position vector of a general point on which is $\mathbf{R}$, the equation of the plane will be $(\mathbf{R} - \mathbf{n}).\mathbf{n} = 0$. This plane cuts Γ at points whose parameters s are given by $f(s) = [\mathbf{r}(s) - \mathbf{n}].\mathbf{n} = 0$, and the condition for it to have three-point contact at $s = s_0$ is that $f(s_0) = f'(s_0) = f''(s_0) = 0$. So

$$[\mathbf{r}(s_0) - \mathbf{n}] \, . \, \mathbf{n} = 0, \; \mathbf{r}'(s_0) \, . \, \mathbf{n} = 0, \; \mathbf{r}''(s_0) \, . \, \mathbf{n} = 0.$$

Hence $[\mathbf{R} - \mathbf{r}(s_0)].\mathbf{n}$ is also zero, by subtraction of two of the equations above, and the three vectors

$$[\mathbf{R} - \mathbf{r}(s_0)], \; \mathbf{r}'(s_0), \; \mathbf{r}''(s_0)$$

are all perpendicular to $\mathbf{n}$, and so parallel to the same plane.

Thus

$$[\mathbf{R} - \mathbf{r}(s_0), \mathbf{r}'(s_0), \mathbf{r}''(s_0)] = 0,$$

and this is the equation of the osculating plane at s_0.

Definition 2.13. *The circle drawn with three-point contact with Γ at P is called the* **circle of curvature** *at P, and its centre and radius are called respectively the* **centre** *and* **radius of curvature.**

From the last theorem it is at once evident that this circle lies in the osculating plane and touches Γ at P. As in the case of plane curves, its radius is the reciprocal of κ, the curvature of Γ at P, so we have

Theorem 2.5. *The centre of the circle of curvature at P on Γ lies on the principal normal at distance $1/\kappa$ from P.*

Definition 2.14. *The sphere drawn with four-point contact with Γ at P is called the* **sphere of curvature** *at P, and its centre and radius are called respectively the* **centre** *and* **radius of spherical curvature.**

Some properties of this sphere appear in example 2.10 at the end of this chapter.

The Form of a Curve at any Point

Suppose s is the arc-length measured from some fixed point P on Γ to a neighbouring point Q. Expanding the position vector $\mathbf{R}$ of Q in a Maclaurin series we get

$$\mathbf{R} = \mathbf{r} + s\mathbf{r}' + s^2\mathbf{r}''/2! + s^3\mathbf{r}'''/3! + \ldots,$$

where $\mathbf{r}$ is the position vector of P. Now

$$\mathbf{r}' = \mathbf{t}, \quad \mathbf{r}'' = \kappa\mathbf{p}, \quad \mathbf{r}''' = \kappa'\mathbf{p} + \kappa(-\kappa\mathbf{t} + \lambda\mathbf{b}),$$

where $\mathbf{t}, \mathbf{p}, \mathbf{b}$ is the fundamental triad of vectors at P. So

$$\begin{aligned}\mathbf{R} - \mathbf{r} &= s\mathbf{t} + s^2\kappa\mathbf{p}/2 + s^3(\kappa'\mathbf{p} - \kappa^2\mathbf{t} + \kappa\lambda\mathbf{b})/6 + \ldots \\ &= \mathbf{t}(s - \kappa^2 s^3/6 + \ldots) + \mathbf{p}(\kappa s^2/2 + \kappa' s^3/6 + \ldots) + \\ &\qquad + \mathbf{b}(\kappa\lambda s^3/6 + \ldots).\end{aligned}$$

If x, y, z are the co-ordinates of Q referred to cartesian axes taken along $\mathbf{t}, \mathbf{p}, \mathbf{b}$ then

$$x = s + \ldots, \quad y = \kappa s^2/2 + \ldots, \quad z = \kappa\lambda s^3/6 + \ldots,$$

and so the projections of Γ on the three co-ordinate planes approximate to the shape of the curves

$$2y = \kappa x^2, \quad 2\lambda^2 y^3 = 9\kappa z^2, \quad 6z = \kappa\lambda x^3,$$

as shown in figure 2.3, on the next page.

Expansions for the Tangent, Principal Normal and Binormal at a Neighbouring Point

As in the last section, if $\mathbf{t}, \mathbf{p}, \mathbf{b}$ and $\mathbf{t}_Q, \mathbf{p}_Q, \mathbf{b}_Q$ are the fundamental triads at P and Q, then

$$\mathbf{t}_Q = \mathbf{t} + s\mathbf{t}' + s^2\mathbf{t}''/2 + s^3\mathbf{t}'''/6 + \ldots,$$

where

$$\begin{aligned}\mathbf{t}' &= \kappa\mathbf{p}, \; \mathbf{t}'' = \kappa'\mathbf{p} + \kappa(-\kappa\mathbf{t} + \lambda\mathbf{b}), \\ \mathbf{t}''' &= \kappa''\mathbf{p} + 2\kappa'(-\kappa\mathbf{t} + \lambda\mathbf{b}) + \\ &\qquad + \kappa(-\kappa'\mathbf{t} + \lambda'\mathbf{b} - \kappa^2\mathbf{p} - \lambda^2\mathbf{p}).\end{aligned}$$

So

$$\mathbf{t}_Q = [1 - s^2\kappa^2/2 - s^3\kappa\kappa'/2 + \ldots]\mathbf{t} + \\ + [s\kappa + s^2\kappa'/2 + s^3(\kappa'' - \kappa^3 - \kappa\lambda^2)/6 + \ldots]\mathbf{p} + \\ + [s^2\kappa\lambda/2 + s^3(2\kappa'\lambda + \kappa\lambda')/6 + \ldots]\mathbf{b}.$$

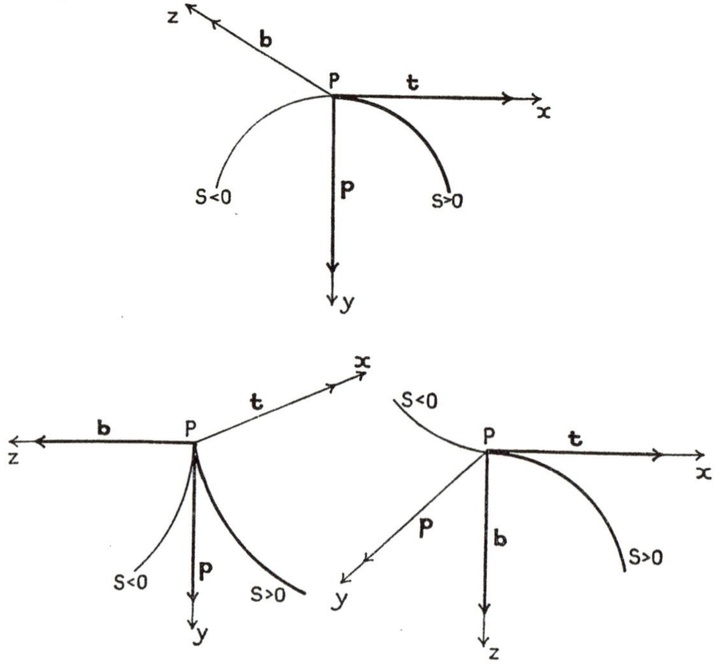

FIG. 2.3

Similarly we can show that

$$\mathbf{p}_Q = [-s\kappa - s^2\kappa'/2 + s^3(-\kappa'' + \kappa^3 + \kappa\lambda^2)/6 + \ldots]\mathbf{t} + \\ + [1 - s^2(\kappa^2 + \lambda^2)/2 + \\ + s^3(-3\kappa\kappa' - 3\lambda\lambda')/6 + \ldots]\mathbf{p} + \\ + [s\lambda + s^2\lambda'/2 + s^3(-\kappa^2\lambda + \lambda'' - \lambda^3)/6 + \ldots]\mathbf{b}$$

and $\mathbf{b}_Q$ may be obtained as $\mathbf{t}_Q \wedge \mathbf{p}_Q$.

The Shortest Distance Between Neighbouring Tangents

The shortest distance d between the tangents $\mathbf{t}$ and $\mathbf{t}_Q$ at P and Q lies in a direction perpendicular to both, that is in the direction of the vector $\mathbf{t} \wedge \mathbf{t}_Q$. We can find d by projecting the length PQ on a vector in this direction. So the shortest distance is

$$|(\mathbf{r}_Q - \mathbf{r}) . (\mathbf{t} \wedge \mathbf{t}_Q)|/|\mathbf{t} \wedge \mathbf{t}_Q|.$$

Now

$$\mathbf{t} \wedge \mathbf{t}_Q = [-s^2\kappa\lambda/2 - s^3(2\kappa'\lambda + \kappa\lambda')/6 + \ldots]\mathbf{p} + \\ + [s\kappa + s^2\kappa'/2 + s^3(\kappa'' - \kappa^3 - \kappa\lambda^2)/6 + \ldots]\mathbf{b},$$

and

$$\mathbf{r}_Q - \mathbf{r} = [s - s^3\kappa^2/6 + \ldots]\mathbf{t} + \\ + [s^2\kappa/2 + s^3\kappa'/6 + \ldots]\mathbf{p} + [s^3\kappa\lambda/6 + \ldots]\mathbf{b}.$$

So

$$d = \frac{|(\mathbf{r}_Q - \mathbf{r}) . (\mathbf{t} \wedge \mathbf{t}_Q)|}{|\mathbf{t} \wedge \mathbf{t}_Q|} = \frac{|(-\kappa^2\lambda/4 + \kappa^2\lambda/6)s^4 + \ldots|}{|\kappa s + \ldots|} \\ = |-\kappa\lambda s^3|/12 + \ldots,$$

where the remaining terms are of degree four at least in s.

In the same way it can be shown that the normals $\mathbf{p}$, $\mathbf{p}_Q$ are at a distance apart of the order of s.

These two results have a bearing on the fact, which will be discussed later, that the tangents to a twisted curve form a developable surface, whereas the normals, in general, do not. One can say, loosely, that 'consecutive' tangents to a twisted curve 'intersect', and that the 'plane' they determine is the osculating plane, whereas 'consecutive' normals do not intersect. Stated precisely, these facts may be put in the form of

Theorem 2.6.

(i) $\lim_{\delta s \to 0} [\delta\mathbf{r}, \mathbf{t}, \mathbf{t} + \delta\mathbf{t}]/\delta s^2 = 0;$

(ii) $\lim_{\delta s \to 0} [\delta\mathbf{r}, \mathbf{p}, \mathbf{p} + \delta\mathbf{p}]/\delta s^2 \neq 0.$

(i) The first limit is $\lim_{\delta s \to 0} [\delta \mathbf{r}, \mathbf{t}, \delta \mathbf{t}]/\delta s^2$, since $\mathbf{t} \wedge \mathbf{t} = \mathbf{0}$, and this is

$$\lim_{\delta s \to 0} \left[\frac{\delta \mathbf{r}}{\delta s}, \mathbf{t}, \frac{\delta \mathbf{t}}{\delta s}\right] = [\mathbf{t},\mathbf{t},\mathbf{p}] = 0.$$

(ii) The second limit is $\lim_{\delta s \to 0} [\delta \mathbf{r}, \mathbf{p}, \delta \mathbf{p}]/\delta s^2$, since $\mathbf{p} \wedge \mathbf{p} = \mathbf{0}$, and this is

$$\lim_{\delta s \to 0} \left[\frac{\delta \mathbf{r}}{\delta s}, \mathbf{p}, \frac{\delta \mathbf{p}}{\delta s}\right] = [\mathbf{t},\mathbf{p},(-\kappa \mathbf{t} + \lambda \mathbf{b})]$$

$$= [\mathbf{t},\mathbf{p},\lambda \mathbf{b}] = \lambda,$$

and it is to be remembered that the condition for three vectors **a**, **b**, **c** to be parallel to the same plane is $[\mathbf{a},\mathbf{b},\mathbf{c}] = 0$.

Theorem 2.7. *If A is a fixed point on a curve Γ and P a point whose arc-distance from A is s, and if A_1 is a fixed point on a second curve Γ_1 and P_1 the point whose arc-distance from A_1 is also s, and if the curvature and torsion at P are respectively equal to the curvature and torsion at P_1 for all values of s, then Γ, Γ_1 are identical in shape.*

For consider Γ fixed and Γ_1 oriented so that A_1 coincides with A, and $\mathbf{t}_1$, $\mathbf{p}_1$, $\mathbf{b}_1$ at A_1 on Γ_1 coincide with **t**, **p**, **b** at A on Γ. Then, for every pair of points P, P_1,

$$\frac{d}{ds}(\mathbf{t}.\mathbf{t}_1 + \mathbf{p}.\mathbf{p}_1 + \mathbf{b}.\mathbf{b}_1) = \kappa \mathbf{p}.\mathbf{t}_1 + \mathbf{t}.\kappa_1 \mathbf{p}_1 + (-\kappa \mathbf{t} + \lambda \mathbf{b}).\mathbf{p}_1 +$$
$$+ \mathbf{p}.(-\kappa_1 \mathbf{t}_1 + \lambda_1 \mathbf{b}_1) - \lambda \mathbf{p}.\mathbf{b}_1 + \mathbf{b}.(-\lambda_1 \mathbf{p}_1)$$
$$= 0.$$

So $(\mathbf{t}.\mathbf{t}_1 + \mathbf{p}.\mathbf{p}_1 + \mathbf{b}.\mathbf{b}_1)$ is a constant, and taking the value at $A = A_1$, where $\mathbf{t}.\mathbf{t}_1 = \mathbf{p}.\mathbf{p}_1 = \mathbf{b}.\mathbf{b}_1 = 1$, this constant has the value 3. But for any two vectors **a**, **b** we have $\mathbf{a}.\mathbf{b} = ab \cos \theta$, so $\mathbf{a}.\mathbf{b} \leqslant ab$. Now all six vectors above are of unit length, so that the maximum value of any product such as $\mathbf{t}.\mathbf{t}_1$ is unity.

Thus $(\mathbf{t}.\mathbf{t}_1 + \mathbf{p}.\mathbf{p}_1 + \mathbf{b}.\mathbf{b}_1) = 3$ implies that each scalar product is unity, and so, for all values of s we get $\mathbf{t} = \mathbf{t}_1$, $\mathbf{p} = \mathbf{p}_1$, $\mathbf{b} = \mathbf{b}_1$.

With Γ_1 oriented as indicated the two curves therefore coincide, and so Γ_1 and Γ are identical in shape.

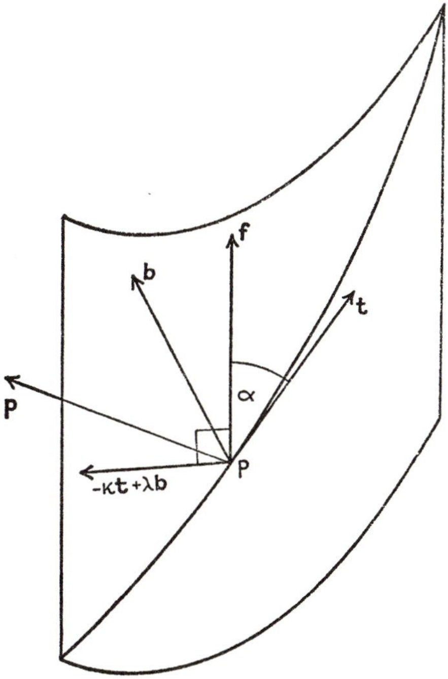

FIG. 2.4

Helices

A simple class of twisted curves is the class of helices, and a particular type of helix — the circular helix — is the curve traced by moving a point along the groove of a uniform screw thread. The same type of curve is formed by a uniformly coiled longitudinal spring.

Definition 2.15. *A twisted curve with the property that the tangents at all points are inclined at the same angle to a fixed direction is called a* **helix.**

If we draw straight lines parallel to the fixed direction through all the points of the curve they will generate a cylinder, not necessarily circular, on which the helix lies.

Let **f** be a unit vector in the fixed direction. Then, **t** being the tangent at P, we are given that

$$\mathbf{f}.\mathbf{t} = \cos\alpha, \text{ (see figure 2.4.)}$$

where α is constant. Differentiating with respect to arc we get $\mathbf{f}.\kappa\mathbf{p} = 0$, so that **p** is perpendicular to **f**, so **f** must be of the form

$$\mathbf{f} = \rho\mathbf{t} + \sigma\mathbf{b},$$

so

$$\cos\alpha = \mathbf{f}.\mathbf{t} = \rho.$$

Since **f** is a unit vector $\rho^2 + \sigma^2 = 1$, so $\sigma = \pm\sin\alpha$. Also, from $\mathbf{f}.\mathbf{p} = 0$, by differentiating again,

$$\mathbf{f}.(-\kappa\mathbf{t} + \lambda\mathbf{b}) = 0,$$

and substituting for **f**

$$(\mathbf{t}\cos\alpha \pm \mathbf{b}\sin\alpha).(-\kappa\mathbf{t} + \lambda\mathbf{b}) = 0,$$

so $(-\kappa\cos\alpha \pm \lambda\sin\alpha) = 0$, and hence $\kappa/\lambda = \pm\tan\alpha$, and we have proved

Theorem 2.8. *For a helix the ratio of curvature to torsion is constant.*

The converse is also true. For suppose that $\kappa/\lambda = \tan\alpha$, where $\tan\alpha$ is constant. Then, from $\mathbf{t}' = \kappa\mathbf{p}$, and $\mathbf{b}' = -\lambda\mathbf{p}$, we get $\mathbf{t}'\cos\alpha + \mathbf{b}'\sin\alpha = \mathbf{0}$, and integrating,

$$\mathbf{t}\cos\alpha + \mathbf{b}\sin\alpha = \text{a constant vector} = \mathbf{u}, \text{ say},$$

where **u** is a unit vector, since the vector on the left of the equation is a unit vector. Taking the scalar product of **t** and **u**

$$\mathbf{t}.\mathbf{u} = \mathbf{t}.\mathbf{t}\cos\alpha + \mathbf{t}.\mathbf{b}\sin\alpha = \cos\alpha,$$

since $\mathbf{t}.\mathbf{t} = 1$ and $\mathbf{t}.\mathbf{b} = 0$. This shows that **t** makes a constant angle with **u**, and so the curve is a helix.

There are other interesting properties to notice. First, the principal normals to a helix are all perpendicular to the fixed direction, and they are also normals to the cylinder on which the helix is drawn. Second, if the cylinder is developed into a plane by rolling it out flat, the helix develops into a straight line. Now a straight line is the shortest distance between two points in a plane, and so the helix is the shortest path on the surface of the cylinder from one point to another. This property of a curve gives it the name of a geodesic. (Another example of a geodesic is a great circle on the surface of a sphere.)

These two properties, that the helix is a geodesic on the cylinder, and that its principal normals are normals to the cylinder, are dependent on one another. For to obtain geodesics on a surface we might draw tight a string over the surface to get the shortest path between two given points. Assuming smooth contact, the reactions on the string are bound to be perpendicular both to the surface and to the string, and this implies coincidence of the normals, since at any point the osculating plane for the string will contain both the principal normal to the string and the normal to the surface.

Involutes and Evolutes

Definition 2.16. *The orthogonal trajectories of the tangents to a curve Γ are called the* **involutes** *of Γ. Γ itself is called an* **evolute** *of any such orthogonal trajectory.*

This definition means that an involute of Γ is a curve in space which cuts every tangent to Γ in such a way that its own tangent at the point of intersection is perpendicular to the tangent to Γ. A practical method of describing an involute of Γ is to imagine a piece of string wrapped along the length of Γ and fastened to some point of Γ at one end. If the other end is now pulled away from Γ, the string always remaining taut, with the straight portion tangent to Γ, this free end will now describe an involute of Γ, and the different involutes of Γ are obtained by altering the length of the string.

Relations between the Corresponding Elements of a Curve Γ and One of its Involutes Γ_1

Suppose P is a point on Γ, and P_1 the point on the tangent $\mathbf{t}$ to Γ at P where it is cut by Γ_1. Let $PP_1 = u$, where u is positive or negative according as PP_1 is in the same direction as $\mathbf{t}$, or in the opposite direction. From our definition the tangent $\mathbf{t}_1$ to Γ_1 at P_1 is perpendicular to $\mathbf{t}$, and so first we obtain an expression for $\mathbf{t}_1$. Throughout we shall use the suffix 1 to indicate elements of Γ_1 at P_1 corresponding to those of Γ at P. Then, with obvious notation,

$$\mathbf{r}_1 = \mathbf{r} + u\mathbf{t},$$

so, as

$$\mathbf{t}_1 = d\mathbf{r}_1/ds_1 = \mathbf{r}_1',$$

$$\mathbf{t}_1 = (\mathbf{r}' + u'\mathbf{t} + u\mathbf{t}')\, ds/ds_1,$$

$$= (\mathbf{t} + u'\mathbf{t} + u\kappa\mathbf{p})\, ds/ds_1.$$

But $\mathbf{t}_1$ is perpendicular to $\mathbf{t}$, so $\mathbf{t}_1.\mathbf{t} = 0$, and therefore $0 = \mathbf{t}.(\mathbf{t} + u'\mathbf{t} + u\kappa\mathbf{p})\, ds/ds_1$, giving $0 = (1 + u')\, ds/ds_1$, since $\mathbf{t}.\mathbf{t} = 1$, $\mathbf{t}.\mathbf{p} = 0$. Assuming $ds/ds_1 \neq 0$ (equality would mean s constant) $(1 + u') = 0$, and so $(s + u) = c$, where c is a constant.

This shows the property connected with unwinding a string from Γ, for it proves that the distance along the curve Γ from a fixed point on Γ to P together with the distance PP_1 is of constant length.

We now investigate the curvature and torsion of Γ_1. From above, since $(1 + u') = 0$,

$$\mathbf{t}_1 = u\kappa\mathbf{p}\, ds/ds_1.$$

This shows that $\mathbf{t}_1 = \pm\,\mathbf{p}$, and $ds/ds_1 = \pm\, 1/u\kappa$. We have agreed earlier that κ shall be essentially positive, and we now choose the sense of measurement for s_1 so that ds/ds_1 is positive when u is positive, and by taking c sufficiently large, we can ensure that u is positive for all relevant values of s and s_1. So

$$\mathbf{t}_1 = \mathbf{p} \quad \text{and} \quad ds_1/ds = \kappa(c - s).$$

So the tangent to Γ_1 at P_1 is parallel to the principal normal to Γ at P.

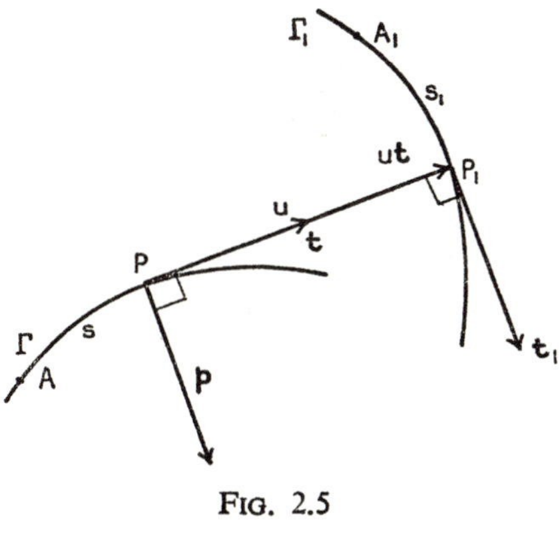

FIG. 2.5

Continuing,

$$\frac{d}{ds_1}(\mathbf{t}_1) = \mathbf{p}'\frac{ds}{ds_1},$$

so

$$\kappa_1\mathbf{p}_1 = (-\,\kappa\mathbf{t} + \lambda\mathbf{b})\, ds/ds_1,$$

and, as $\mathbf{p}_1$ is a unit vector,

$$\mathbf{p}_1 = (-\,\kappa\mathbf{t} + \lambda\mathbf{b})/(\kappa^2 + \lambda^2)^{1/2}.$$

Hence

$$\kappa_1^2(ds_1/ds)^2 = \kappa^2 + \lambda^2,$$

and

$$\kappa_1 = (\kappa^2 + \lambda^2)^{1/2}/\kappa(c - s).$$

Now

$$\mathbf{b}_1 = \mathbf{t}_1 \wedge \mathbf{p}_1 = \mathbf{p} \wedge (-\kappa\mathbf{t} + \lambda\mathbf{b})(1/\kappa_1)\, ds/ds_1$$
$$= (\kappa\mathbf{b} + \lambda\mathbf{t})/(\kappa^2 + \lambda^2)^{1/2}.$$

So

$$-\lambda_1\mathbf{p}_1\, ds_1/ds$$
$$= \frac{\kappa'\mathbf{b} - \kappa\lambda\mathbf{p} + \lambda'\mathbf{t} + \lambda\kappa\mathbf{p}}{(\kappa^2 + \lambda^2)^{1/2}} - \frac{(\kappa\kappa' + \lambda\lambda')(\kappa\mathbf{b} + \lambda\mathbf{t})}{(\kappa^2 + \lambda^2)^{3/2}}$$
$$= \frac{(\kappa^2 + \lambda^2)(\kappa'\mathbf{b} + \lambda'\mathbf{t}) - (\kappa\kappa' + \lambda\lambda')(\kappa\mathbf{b} + \lambda\mathbf{t})}{(\kappa^2 + \lambda^2)^{3/2}}$$
$$= \frac{(\kappa\lambda' - \kappa'\lambda)(\kappa\mathbf{t} - \lambda\mathbf{b})}{(\kappa^2 + \lambda^2)^{3/2}}.$$

Hence

$$\lambda_1 = (\kappa\lambda' - \kappa'\lambda)/(\kappa^2 + \lambda^2)(ds_1/ds)$$
$$= (\kappa\lambda' - \kappa'\lambda)/(\kappa^2 + \lambda^2)\kappa(c - s).$$

Collecting these results we have $ds_1/ds = \kappa(c - s)$, and

$$\mathbf{t}_1 = \mathbf{p},$$
$$\mathbf{p}_1 = (-\kappa\mathbf{t} + \lambda\mathbf{b})/(\kappa^2 + \lambda^2)^{1/2},$$
$$\mathbf{b}_1 = (\lambda\mathbf{t} + \kappa\mathbf{b})/(\kappa^2 + \lambda^2)^{1/2},$$
$$\kappa_1 = (\kappa^2 + \lambda^2)^{1/2}/\kappa(c - s),$$
$$\lambda_1 = (\kappa\lambda' - \kappa'\lambda)/\kappa(\kappa^2 + \lambda^2)(c - s).$$

The Evolutes of a Curve Γ

If Γ_1 is an evolute of a curve Γ, then, from our definition, Γ is an orthogonal trajectory of the tangents to Γ_1. So the tangent at P to Γ is perpendicular to the tangent at the corresponding point P_1 on Γ_1, and P_1 must lie in the normal plane of Γ at P. So

$$\mathbf{r}_1 = \mathbf{r} + u\mathbf{p} + v\mathbf{b},$$

where u and v are to be determined. From this we get

$$\mathbf{t}_1 = [\mathbf{t} + u'\mathbf{p} + u(-\kappa\mathbf{t} + \lambda\mathbf{b}) + v'\mathbf{b} + v(-\lambda\mathbf{p})](ds/ds_1),$$

and this is a vector proportional to the vector $(u\mathbf{p} + v\mathbf{b})$. So

$$1 - u\kappa = 0 \ , \ (u' - \lambda v)(ds/ds_1) = \rho u,$$

and

$$(\lambda u + v')(ds/ds_1) = \rho v.$$

From these relations $u = 1/\kappa$, and $v(u' - \lambda v) = u(\lambda u + v')$.

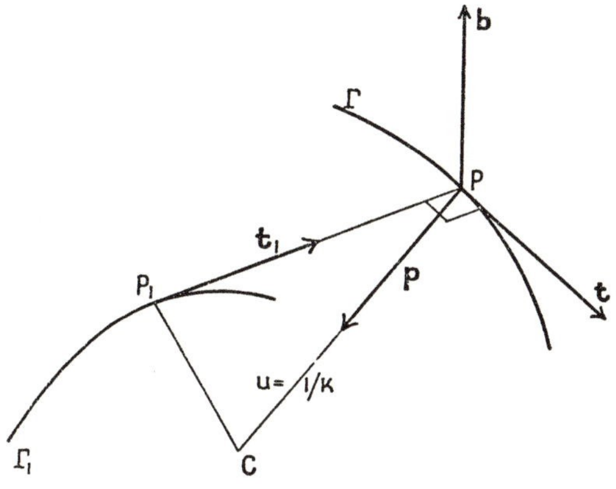

FIG. 2.6

The last equation gives

$$\frac{u'v - uv'}{v^2} = \frac{\lambda(u^2 + v^2)}{v^2} = \lambda\left[\frac{u^2}{v^2} + 1\right].$$

So

$$\frac{d}{ds}\left[\frac{u}{v}\right] = \lambda(u^2 + v^2)/v^2,$$

and

$$\int \frac{d(u/v)}{1 + (u^2/v^2)} = \int \lambda\, ds,$$

giving $\tan^{-1}(u/v) = \int \lambda\, ds + \text{constant} = \Lambda + c$, say, where $\Lambda = \int \lambda\, ds$. So $1/\kappa = v \tan(\Lambda + c)$, and the point P_1 is given by

$$\mathbf{r}_1 = \mathbf{r} + [\mathbf{p} + \cot(\Lambda + c)\mathbf{b}]/\kappa.$$

We see from this that the curve Γ has an infinity of evolutes obtained by letting c take all values.

If Γ is a plane curve, then $\lambda = 0$, and so the evolutes of Γ are given by $u/v = \text{constant}$, and are

$$\mathbf{r}_1 = \mathbf{r} + \frac{1}{\kappa}(\mathbf{p} + \omega\mathbf{b}),$$

where ω is a constant whose different values determine the different evolutes.

Now in the first chapter dealing with plane curves we talked of the evolute of a curve, meaning a curve lying in the same plane. We now see that a plane curve Γ has an infinity of evolutes, each of which lies on a cylinder drawn perpendicular to the plane of Γ, and whose section by this plane is the evolute $\mathbf{r}_1 = \mathbf{r} + \mathbf{p}/\kappa$, the locus of the centres of curvature of Γ.

Furthermore, the tangent to such an evolute is

$$\begin{aligned}\mathbf{t}_1 &= d\mathbf{r}_1/ds_1 \\ &= \left[d\mathbf{r}/ds + \frac{1}{\kappa}(-\kappa\mathbf{t} + \lambda\mathbf{b} - \omega\lambda\mathbf{p}) - \frac{\kappa'}{\kappa^2}(\mathbf{p} + \omega\mathbf{b})\right](ds/ds_1),\end{aligned}$$

which, since $\lambda = 0$ and $d\mathbf{r}/ds = \mathbf{t}$, reduces to the vector

$$\pm(\mathbf{p} + \omega\mathbf{b})/(1 + \omega^2)^{1/2},$$

since $\mathbf{t}_1$ is a unit vector. And this vector makes a constant angle with $\mathbf{b}$, which itself is fixed in direction for a plane curve. So we get

Theorem 2.9. *The evolutes of a plane curve are helices.*

Related Curves

The work above on involutes and evolutes is a special type of the work on pairs of curves whose points, or elements such as the tangent, principal normal, and binormal, are related in some special way, and the methods used above may be applied equally well in such cases. Some examples of such relations will be found at the end of the chapter.

EXAMPLES ON CHAPTER TWO

2.1. Find the unit tangent vector $\mathbf{t}$ at the point θ on the twisted cubic curve whose position vector $\mathbf{r}$ is given by $\mathbf{r} = \mathbf{i}a\theta^3 + \mathbf{j}b\theta^2 + \mathbf{k}c\theta$, and write down its cartesian equations.

2.2. For a curve Γ the position vector $\mathbf{r}$ satisfies the relation $\ddot{\mathbf{r}} = \mathbf{0}$. Show that Γ is either a straight line or a single point.

2.3. Find the curves which can be represented by $\dddot{\mathbf{r}} = \mathbf{0}$.

2.4. Find $\mathbf{t}$, $\mathbf{p}$, $\mathbf{b}$, κ in terms of a general parameter t.

2.5. If Γ is $\mathbf{r} = \mathbf{r}(t)$ prove that $\kappa^2\lambda = [\dot{\mathbf{r}},\ddot{\mathbf{r}},\dddot{\mathbf{r}}]/\dot{s}^6$.

2.6. Find κ, λ for the curves

(i) $\mathbf{r} = \mathbf{i}a(3t - t^3) + \mathbf{j}a(3t^2) + \mathbf{k}a(3t + t^3)$;

(ii) $\mathbf{r} = \mathbf{i}a \cosh t + \mathbf{j}a \sinh t + \mathbf{k}ct$;

(iii) $\mathbf{r} = \mathbf{i}a(\theta - \sin\theta) + \mathbf{j}a(1 - \cos\theta) + \mathbf{k}b\theta$.

2.7. If $\mathbf{u}$, $\mathbf{v}$, $\mathbf{w}$ are three mutually perpendicular vectors, each of unit length, and each a function of a single parameter t, show that $\dot{\mathbf{u}}$, $\dot{\mathbf{v}}$, $\dot{\mathbf{w}}$ are all parallel to the same plane.

2.8. Prove that $\mathbf{t}'.\mathbf{p}' = 0$, $\mathbf{p}'.\mathbf{b}' = 0$, $\mathbf{b}'.\mathbf{t}' = -\kappa\lambda$, and show that $\mathbf{t}'$, $\mathbf{p}'$, $\mathbf{b}'$ are all perpendicular to the Darboux vector $\lambda\mathbf{t} + \kappa\mathbf{b}$.

2.9. Find cartesian equations for the three principal planes at $\mathbf{r} = \mathbf{r}(t)$ on a curve Γ.

2.10. Show that the position vector of the centre of spherical curvature is $\mathbf{r} + (\mathbf{p}/\kappa) - (\mathbf{b}\kappa'/\kappa^2\lambda)$.

If R is the radius of spherical curvature show that

$$R = |(\mathbf{t} \wedge \mathbf{t}'')/\kappa^2\lambda|.$$

If ρ and σ are respectively the radius of curvature and torsion for a curve, prove that

$$R^2 = \rho^2 + (\sigma\rho')^2.$$

Find conditions for

(i) The centre of spherical curvature to coincide with the centre of curvature;

(ii) R to be constant.

Show that for a curve lying on the surface of a sphere

$$\lambda/\kappa = \frac{d}{ds}(\kappa'/\kappa^2\lambda).$$

2.11. Show that the normals $\mathbf{p}$, $\mathbf{p} + \delta\mathbf{p}$ are approximately a distance $\lambda\,\delta s(\kappa^2 + \lambda^2)^{-1/2}$ apart, and that the angle between them is approximately $\delta s(\kappa^2 + \lambda^2)^{1/2}$.

2.12. Show that necessary and sufficient conditions for a curve to be a helix is that the principal normals should all be parallel to a fixed plane.

2.13. For the helix $\mathbf{r} = \mathbf{i}a\cos\theta + \mathbf{j}a\sin\theta + \mathbf{k}a\,\theta\tan\alpha$ drawn on a circular cylinder show that κ and λ are both constant.

For any curve prove that $[\mathbf{t}',\mathbf{t}'',\mathbf{t}'''] = \kappa^5(\lambda/\kappa)'$, and show that a necessary and sufficient condition for a curve to be a helix is that $[\mathbf{r}'',\mathbf{r}''',\mathbf{r}''''] = 0$.

2.14. Prove that the locus of the centre of curvature of a helix lying on a circular cylinder is also a helix, and find the condition for it to lie on the same cylinder.

2.15. A curve is drawn on the surface of a right circular cone so as to cut all the generators of the cone at the same angle. Show that its orthogonal projection on the base of the cone is an equiangular spiral.

2.16. Two curves Γ, Γ_1 are such that the binormals to Γ_1 are principal normals to Γ.

(i) Find $\mathbf{t}_1$;

(ii) Show that the corresponding points of Γ, Γ_1 are a constant distance c apart;

(iii) Prove that $c(\kappa^2 + \lambda^2) = \kappa$.

2.17. A point P_1 is taken on the tangent $\mathbf{t}$ at the point P on a curve Γ, so that $PP_1 = c$, a constant. Prove that $\mathbf{t}_1$ is parallel to the osculating plane of Γ at P, and find necessary and sufficient conditions for the locus of P_1 to be a straight line.

2.18. Two curves Γ, Γ_1 are so related that the principal normal to Γ at P lies along the principal normal to Γ_1 at P_1 for all pairs of points P, P_1. Show that the tangents at P and P_1 are inclined at a constant angle.

Show also that κ and λ at P on Γ satisfy the same linear relation for all points P, and that κ_1, λ_1 at P_1 on Γ_1 are related to κ, λ by the equation

$$a^2\lambda\lambda_1 + a(\kappa_1 - \kappa) - a^2\kappa\kappa_1 = 0,$$

where a is a constant whose geometrical significance is to be determined.

2.19. Show that if the involutes of a twisted curve are plane curves then the curve is a helix.

2.20. The evolutes of a curve Γ cut one of its normal planes in points P_1. Show that all such points P_1 lie on a straight line, parallel to the binormal at P, which passes through C, the centre of curvature of Γ at P.

CHAPTER THREE

Surfaces

Introduction

Perhaps the most natural starting point for discussing surfaces is to consider the equation $z = f(x,y)$ in three-dimensional Euclidean space. For if we take the plane $z = 0$ to be horizontal then $f(x,y)$ can be regarded as the height of a point P on the surface lying vertically above the point Q, on $z = 0$, whose coordinates are (x,y), and, as Q moves over a region on $z = 0$, so P will move over a corresponding region on the surface $z = f(x,y)$. Thinking in geographical terms, the curve given by sectioning the surface with a plane $z = k$ is then a contour line.

Difficulties arise, however, as surfaces are frequently given in the form $g(x,y,z) = 0$; for example, the equation of a sphere of radius c, whose centre is the origin 0, is $x^2 + y^2 + z^2 = c^2$. In this case $z = \pm(c^2 - x^2 - y^2)^{1/2}$, and we get two single-valued functions z by taking the separate signs. Each of these represents one-half of the surface of the sphere. The contour lines would then be circles given by taking values of k lying between $+ c$ and $- c$ in the pair of equations $z = k$, $x^2 + y^2 = c^2 - k^2$.

Also, while it appears intuitive that an equation $g(x,y,z) = 0$ is a two-dimensional locus, that is, a surface, since only two of the variables x, y, z can be arbitrarily assigned, the problem of expressing this surface by an equation of the form $z = f(x,y)$ presents difficulties.

Similarly, when two equations $g(x,y,z) = 0$, $h(x,y,z) = 0$ are given, representing two surfaces, it would appear that for points common to both surfaces only one of the variables can be regarded as arbitrary; so that such points would seem to form a one-dimensional locus, that is, a curve. But again, the problem of determining such curves of intersection of surfaces is not in general by any means simple.

To avoid these difficulties we shall adopt a parametric form of representation for the coordinates of a point on a surface, which, using cartesian coordinates, will take the form $x = x(u,v)$, $y = y(u,v)$, $z = z(u,v)$, where u, v are our parameters, and where the functions of them are subject to certain restrictions. If the parameters u, v are themselves functions of a third parameter t, then x, y, z become functions of this single parameter, and the locus of $P(x,y,z)$ becomes a curve.

For points on a sphere we obtain parametric coordinates as follows. Using geographical terms for a sphere whose centre is O and whose radius is c, take axes Ox, Oy in the plane of the equator and take the axis Oz along the radius through the north pole N_z. Let P be any point on the surface of the sphere and let the meridian through P make an angle ϕ with the meridian which cuts the axis Ox. Let θ be the angle between OP and ON_z, and let M and N be respectively the projections of P on the axis of Oz and on the plane Oxy. Then we have

$$\begin{aligned} PM &= c \sin \theta = ON \\ x &= ON \cos \phi = c \sin \theta \cos \phi \\ y &= ON \sin \phi = c \sin \theta \sin \phi \\ z &= c \cos \theta. \quad \text{(See figure 3.1.)} \end{aligned}$$

As P moves from N_z to S along the meridian N_zPS, θ ranges from 0 to π. As the meridian through P moves round the equator from N_zAS back to N_zAS, ϕ ranges from 0 to 2π.

If we make θ and ϕ depend on some further parameter t, so that $\theta = \theta(t)$, $\phi = \phi(t)$, then P will describe a curve on the sphere. In particular, taking θ constant gives the parallels of latitude, while taking ϕ constant gives the meridians.

For a cylinder we have $z = u$, $x = r\cos\theta$, $y = r\sin\theta$; for a cone $x = u\tan\alpha\cos\phi$, $y = u\tan\alpha\sin\phi$, $z = u$; and for a surface of revolution $x = r\cos\theta$, $y = r\sin\theta$, $z = f(r)$; as can be seen from figure 3.2.

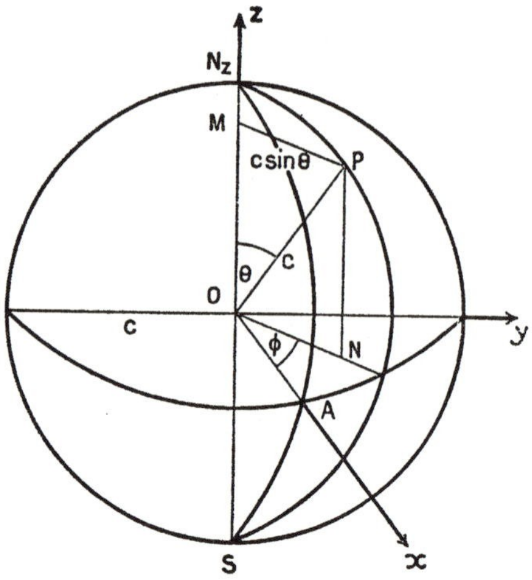

FIG. 3.1

These ideas of parametric representation we shall now express in terms of the position vector $\mathbf{r}$ of the point P, and give

Definition 3.1. *Suppose* $\mathbf{r}$, *the position vector of a point P in three-dimensional Euclidean space, is a function of two real parameters u and v, then the locus of P as u and v vary will be called a* **surface** *S.*

We shall assume that the function $\mathbf{r} = \mathbf{r}(u,v)$ satisfies the following conditions at almost all points (that is at all points corresponding to points (u,v) in the u-v plane, with the possible

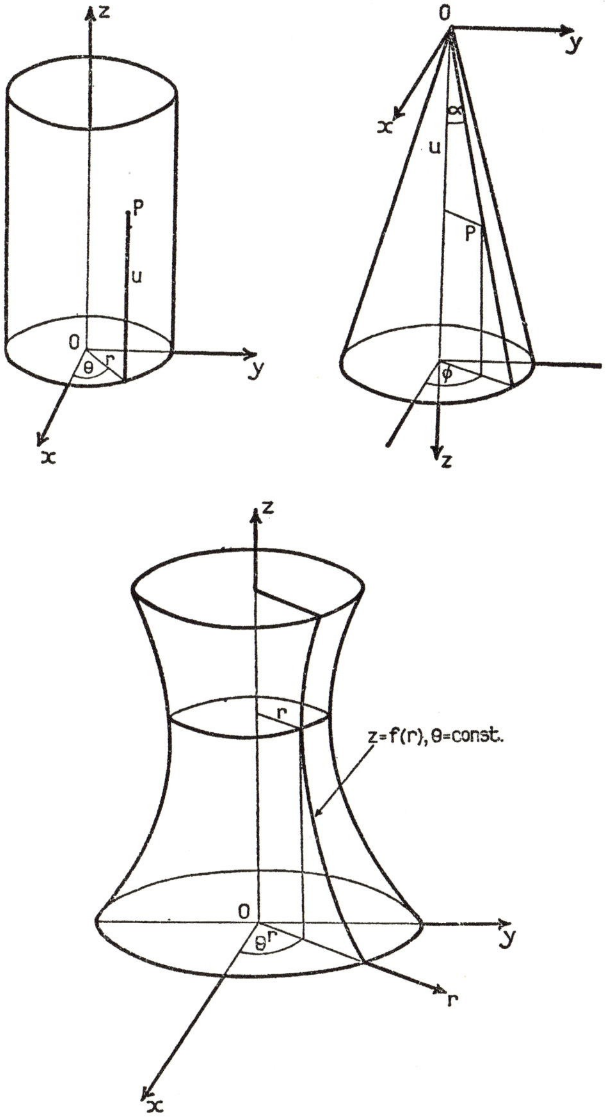

FIG. 3.2

exception of points lying on a finite number of curves in this plane):

(a) $\mathbf{r}(u,v)$ is single-valued;

(b) $\mathbf{r}(u,v)$ possesses continuous partial derivatives up to any order required;

(c) $\dfrac{\partial \mathbf{r}}{\partial u} \wedge \dfrac{\partial \mathbf{r}}{\partial v} \neq \mathbf{0}$.

With these conditions, and the usual cartesian coordinates, S has a parametric representation

$$x = x(u,v), \quad y = y(u,v), \quad z = z(u,v),$$

from which u and v may be eliminated to obtain the equation of S in the form $f(x,y,z) = 0$.

Notation

We shall write

$$\frac{\partial \mathbf{r}}{\partial u} \quad \mathbf{r}_u, \quad \frac{\partial \mathbf{r}}{\partial v} = \mathbf{r}_v, \quad \frac{\partial^2 \mathbf{r}}{\partial u^2} = \mathbf{r}_{uu}, \quad \frac{\partial^2 \mathbf{r}}{\partial v^2} = \mathbf{r}_{vv},$$

$$\frac{\partial^2 \mathbf{r}}{\partial u \partial v} = \mathbf{r}_{vu} = \mathbf{r}_{uv} = \frac{\partial^2 \mathbf{r}}{\partial v \partial u},$$

the inversion of the order of partial differentiation being allowable in view of the condition (b) above.

Parametric Curves

If one of the parameters v is given a fixed value v_0, then $\mathbf{r}(u,v_0)$ is a function of a single parameter u, and so the locus of P will become a curve on S. Thus $u =$ const., $v =$ const., for different values of the constants, give two families of curves on S with the property that each point $P(u,v)$ on S lies at the intersection of two curves, one from each family.

Definition 3.2. *The curves obtained by making either u or v constant are called* **parametric curves,** *and u and v themselves are called the* **curvilinear coordinates,** *or simply the* **coordinates** *of points on the surface.*

Curves on a Surface

If u, v are themselves single-valued functions of a third parameter t, possessed of derivatives as far as required, then $\mathbf{r}$ becomes a function of the single parameter t, and as t varies the locus of $P(u,v)$ will be a curve Γ on S. The parameters u, v will satisfy an equation $\phi(u,v) = 0$, obtained by eliminating t.

Definition 3.3. *A tangent at any point P of a curve Γ lying on S is called a* **tangent line** *of S.*

Theorem 3.1. *The tangent lines at a point P on S all lie on a plane, called the* **tangent plane** *to S at P.*

Let Γ be a curve on S, and $P(u,v)$ a point on Γ at which du/dt, dv/dt do not vanish simultaneously. Since $\mathbf{r} = \mathbf{r}(u,v) = \mathbf{r}[u(t),v(t)]$,

$$\frac{d\mathbf{r}}{dt} = \mathbf{r}_u \frac{du}{dt} + \mathbf{r}_v \frac{dv}{dt} \neq \mathbf{0}.$$

This shows that $d\mathbf{r}/dt$, a tangent vector to Γ at P, lies in the plane of $\mathbf{r}_u$, $\mathbf{r}_v$, which are tangent vectors to the parametric curves at P. This plane is well defined since $\mathbf{r}_u \wedge \mathbf{r}_v \neq \mathbf{0}$, by condition (c) above. So all the tangent lines to S and P lie in this plane.

Definition 3.4. *The line through P perpendicular to the tangent plane to S at P is called the* **normal** *to S at P.*

Now $\mathbf{r}_u \wedge \mathbf{r}_v$ is a vector perpendicular to the tangent plane, so we define a **unit normal e** by

$$\mathbf{e} = \frac{\mathbf{r}_u \wedge \mathbf{r}_v}{|\mathbf{r}_u \wedge \mathbf{r}_v|}.$$

It is to be noticed that the direction of $\mathbf{e}$ will be reversed if the order of the two vectors on the right is interchanged, and so there are two unit normals at each point of S.

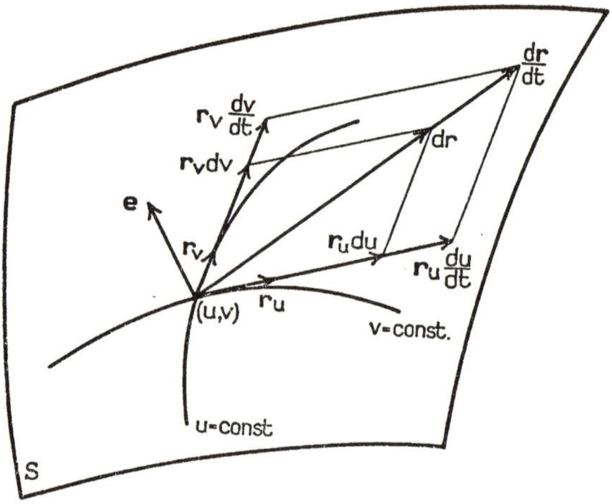

FIG. 3.3

Arc-length of a Curve on a Surface

To define a curve Γ on S, the parameters u, v must satisfy an equation $\phi(u,v) = 0$, being each functions of a parameter t. From theorem 3.1 the tangent vector, $d\mathbf{r}/dt$, to Γ is a linear combination of the vectors $\mathbf{r}_u$ and $\mathbf{r}_v$, and so its direction will be determined by the ratio $\dot{u}/\dot{v}$, or, what is the same thing, $-(\partial\phi/\partial v)/(\partial\phi/\partial u)$, obtained by differentiating $\phi(u,v)$ with respect to t. It will, however, be remembered that in chapter II we showed that $d\mathbf{r} = \dot{\mathbf{r}}\, dt$ was a vector tangential to a curve, and in the same way $\mathbf{r}_u\dot{u}\, dt$ and $\mathbf{r}_v\dot{v}\, dt$ are vectors tangential respectively to the parametric curves $v = \text{const.}$, $u = \text{const.}$, so that, using differentials, and writing $\dot{u}\, dt = du$, and $\dot{v}\, dt = dv$ the vector $d\mathbf{r}$, given by

$$d\mathbf{r} = \mathbf{r}_u\, du + \mathbf{r}_v\, dv$$

is a vector tangential to Γ whose length will depend on the magnitude of dt. It is now evident, that, as u and v are taken respectively as various functions of t, the ratio of the differentials du, dv will determine the direction of Γ. (See figure 3.3.)

From the theory of curves in chapter II the arc-length of a curve is given by $\int ds$, where $ds^2 = (d\mathbf{r})^2 = d\mathbf{r}.d\mathbf{r}$, so for Γ on S we may write

$$\begin{aligned}(d\mathbf{r})^2 = d\mathbf{r}.d\mathbf{r} &= (\mathbf{r}_u\,du + \mathbf{r}_v\,dv).(\mathbf{r}_u\,du + \mathbf{r}_v\,dv)\\ &= \mathbf{r}_u.\mathbf{r}_u\,du^2 + 2\mathbf{r}_u.\mathbf{r}_v\,dudv + \mathbf{r}_v.\mathbf{r}_v\,dv^2,\end{aligned}$$

and the arc-length of Γ will be the integral of the square root of this expression. We now write

$$\mathbf{r}_u.\mathbf{r}_u = E, \quad \mathbf{r}_u.\mathbf{r}_v = F, \quad \mathbf{r}_v.\mathbf{r}_v = G,$$

and give

Definition 3.5. *For given values of the differentials du, dv, defining some curve Γ on S, the* **element of arc** *of Γ is defined as ds, where*

$$ds^2 = E\,du^2 + 2F\,dudv + G\,dv^2.$$

This expression for ds^2 is called the **first fundamental quadratic form** *for the surface.*

If u and v are defined in terms of a parameter t, the arc-length of the corresponding curve Γ lying on S will then be

$$\int [E\dot{u}^2 + 2F\dot{u}\dot{v} + G\dot{v}^2]^{1/2}\,dt.$$

If two curves have the same tangent line at P then the ratio du/dv will be the same for both, and if, in addition, the values of the differentials du, dv are the same for both curves, then the element of arc-length will also be the same.

Area

If $\mathbf{r}_u\,du, \mathbf{r}_v\,dv$, where du, dv are differentials of the parameters u and v, are drawn at P on S they will lie along the tangent lines to the parametric curves at P, and a parallelogram may be drawn with these vectors as two of its sides. Its area, dA, will be $|\mathbf{r}_u\,du \wedge \mathbf{r}_v\,dv|$, and if θ is the angle between the two vectors then

$$\begin{aligned} dA &= |\mathbf{r}_u\,du||\mathbf{r}_v\,dv| \sin\theta \\ &= |\mathbf{r}_u||\mathbf{r}_v||du\,dv\,(1 - \cos^2\theta)^{1/2}| \\ &= [\mathbf{r}_u^2\mathbf{r}_v^2 - (\mathbf{r}_u.\mathbf{r}_v)^2]^{1/2}|dudv| \\ &= (EG - F^2)^{1/2}|dudv|. \end{aligned}$$

The concept of area, like arc-length, is a deep one, and we shall regard it as intuitive that the area dA of this parallelogram is an approximation to the area δS on the surface bounded by the curves $u =$ const., $v =$ const., $u + du =$ const., $v + dv =$ const. We therefore give

Definition 3.6. *The* **element of area** *of S is defined as dS, where*

$$dS = |(EG - F^2)^{1/2}\,dudv|.$$

The area of a finite portion of S is defined to be

$$\iint |(EG - F^2)^{1/2}\,dudv|.$$

Oblique and Normal Sections

Definition 3.7. *Let σ be a plane drawn through a point P on S. Then σ will cut S in a curve Γ, called a* **section** *of S. If σ is drawn to contain the normal to S at P then Γ is called a* **normal section**; *otherwise it is called an* **oblique section.**

OBLIQUE AND NORMAL SECTIONS

Suppose **t**, **p** are the unit tangent and principal normal to Γ at P, and that κ is the curvature of Γ at this point, then with the usual notation,

$$\mathbf{t}' = \kappa\mathbf{p} = \mathbf{r}'' = \frac{d}{ds}(\mathbf{r}') = \frac{d}{ds}(\mathbf{r}_u u' + \mathbf{r}_v v')$$
$$= \mathbf{r}_{uu}(u')^2 + 2\mathbf{r}_{uv}u'v' + \mathbf{r}_{vv}(v')^2 + $$
$$+ \mathbf{r}_u u'' + \mathbf{r}_v v'',$$

and if **e** is one of the unit normals to S at P

$$\kappa\mathbf{p}.\mathbf{e} = \mathbf{e}.(\mathbf{r}_{uu}\, du^2 + 2\mathbf{r}_{uv}\, dudv + \mathbf{r}_{vv}\, dv^2)/ds^2,$$

since $\mathbf{e}.\mathbf{r}_u = 0$ and $\mathbf{e}.\mathbf{r}_v = 0$.

So

$$\kappa\mathbf{p}.\mathbf{e} = (Ldu^2 + 2Mdudv + Ndv^2)/ds^2,$$

where $L = [\mathbf{r}_u,\mathbf{r}_v,\mathbf{r}_{uu}]/W$, $M = [\mathbf{r}_u,\mathbf{r}_v,\mathbf{r}_{uv}]/W$, $N = [\mathbf{r}_u,\mathbf{r}_v,\mathbf{r}_{vv}]/W$, for $\mathbf{e} = (\mathbf{r}_u \wedge \mathbf{r}_v)/|\mathbf{r}_u \wedge \mathbf{r}_v| = (\mathbf{r}_u \wedge \mathbf{r}_v)/(EG - F^2)^{1/2} = (\mathbf{r}_u \wedge \mathbf{r}_v)/W$, where $W = (EG - F^2)^{1/2}$.

Now if θ is the angle between **e** and **p** this gives

$$\kappa\mathbf{p}.\mathbf{e} = \kappa \cos\theta = (Ldu^2 + 2Mdudv + Ndv^2)/ds^2.$$

Now all sections for which the ratio du/dv has the same value have the same tangent **t** at P, and the same value for the last member of the equation, a value which is independent of θ. So $\kappa \cos\theta$ is constant, and in the particular case in which $\theta = 0$ or π we obtain the normal section, and so $\kappa\cos\theta$ and κ_N, the curvature of the normal section, have the same numerical value. Thus we have proved the following theorem:

Theorem 3.2. (Meusnier's theorem). *The curvatures of a normal and an oblique section which have the same tangent line at P on a surface S have values which are numerically in the ratio* $|\cos\theta|:1$, *where* θ *is the angle between their principal normals at P.*

If C, C_N are the centres of curvature of the two sections then $PC_N = 1/\kappa_N$ and $PC = 1/\kappa$, so that $PC_N \cos \theta = PC$, which shows that C lies on a circle of diameter PC_N, lying in the plane through P normal to the tangent at P, which passes through P and has its centre a distance $1/2\kappa_N$ from P along the normal to the surface.

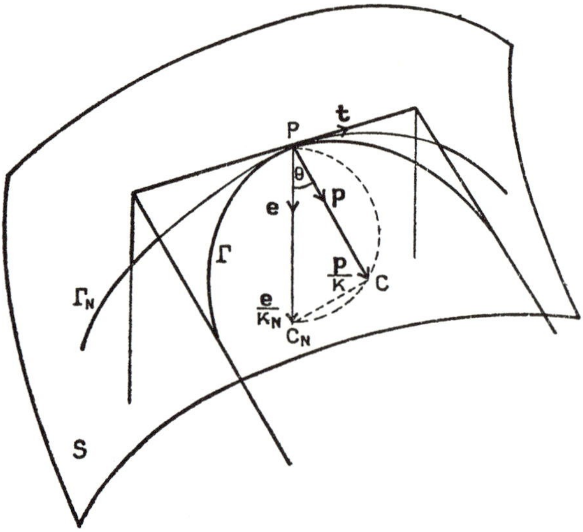

FIG. 3.4

Definition 3.8. *The expression* $Ldu^2 + Mdudv + Ndv^2$ *is called the* **second fundamental quadratic form** *for the surface.*

To determine its significance consider the points P, Q on S, with position vectors $\mathbf{r}$ and $\mathbf{r} + \delta\mathbf{r}$, whose coordinates are (u,v) and $(u + \delta u, v + \delta v)$. Then, if $\mathbf{e}$ is the normal at P, the distance h of Q from the tangent plane is $|PQ \cos \theta|$, where θ is the angle between PQ and $\mathbf{e}$, that is to say $h = |\delta\mathbf{r}.\mathbf{e}|$. But, expanding $\delta\mathbf{r}$ in terms of δu, δv we get

$$\delta\mathbf{r} = (\mathbf{r}_u\delta u + \mathbf{r}_v\delta v) + (\mathbf{r}_{uu}\delta u^2 + 2\mathbf{r}_{uv}\delta u\delta v + \mathbf{r}_{vv}\delta v^2)/2! + \dots$$

assuming that δu, δv are independent, so, as **e** is perpendicular to both $\mathbf{r}_u$ and $\mathbf{r}_v$, giving $\mathbf{e}.\mathbf{r}_u = \mathbf{e}.\mathbf{r}_v = 0$, we get

$$h = |\delta\mathbf{r}.\mathbf{e}| = |(L\delta u^2 + 2M\delta u\delta v + N\delta v^2)/2 + \ldots|.$$

Hence, regarding du, dv as small variations in u and v in passing from P to Q, we obtain the result that

$$|Ldu^2 + 2Mdudv + Ndv^2|$$

is approximately twice the distance h.

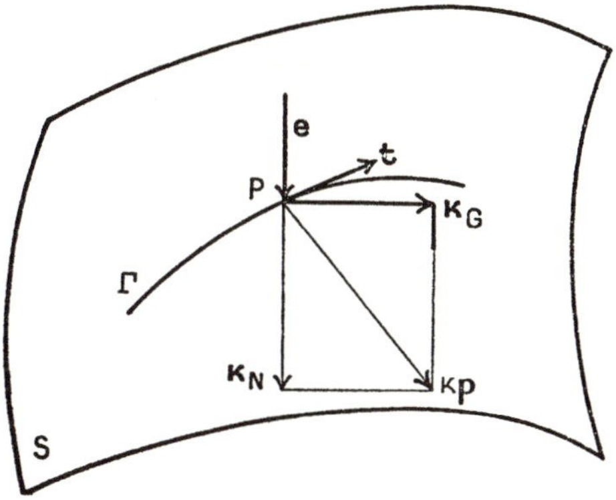

FIG. 3.5

Curvature of Any Curve Lying on a Surface

We now turn our attention to curves lying on a surface which no longer lie in plane sections. If Γ is any such curve of arc-length s whose curvature at P is κ, where $\mathbf{t}' = \kappa\mathbf{p}$, $\mathbf{t}$ being a tangent at P to both Γ and S, and $\mathbf{p}$ the principal normal to Γ, in general not normal to S, we give

Definition 3.9. *The vector* $\boldsymbol{\kappa} = \kappa\mathbf{p}$ *is called the* **curvature vector** *of* Γ.

Now $\boldsymbol{\kappa}$ may be resolved into component vectors $\boldsymbol{\kappa}_N$, $\boldsymbol{\kappa}_G$, respectively normal and tangential to S. We now give

Definition 3.10. $\boldsymbol{\kappa}_N$ *is called the* **normal curvature vector** *and* $\boldsymbol{\kappa}_G$ *is called the* **geodesic curvature vector** *of* Γ.

If we choose our parameters for S so that $\mathbf{e}$, the normal vector to S at P on Γ is not only parallel, but actually in the same direction as $\boldsymbol{\kappa}_N$, then the angle θ between the two vectors $\boldsymbol{\kappa}$ and $\boldsymbol{\kappa}_N$ will be acute, and we shall get $\boldsymbol{\kappa}.\mathbf{e} = \kappa \cos \theta$. But $\mathbf{t}.\mathbf{e} = 0$, so

$$\mathbf{e}.\mathbf{t}' + \mathbf{t}.\mathbf{e}' = \mathbf{e}.\boldsymbol{\kappa} + \mathbf{t}.\mathbf{e}' = 0.$$

This gives $\kappa \cos \theta = -\,\mathbf{t}.\mathbf{e}' = -\,\mathbf{r}'.\mathbf{e}'$, which, in terms of parameters, becomes

$$\begin{aligned}\kappa \cos \theta &= -\frac{(\mathbf{r}_u du + \mathbf{r}_v dv).(\mathbf{e}_u du + \mathbf{e}_v dv)}{ds^2} \\ &= -\frac{\mathbf{r}_u.\mathbf{e}_u du^2 + (\mathbf{r}_u.\mathbf{e}_v + \mathbf{r}_v.\mathbf{e}_u) du dv + \mathbf{r}_v.\mathbf{e}_v dv^2}{ds^2}\end{aligned}$$

Now from $\mathbf{r}_u.\mathbf{e} = 0$ and $\mathbf{r}_v.\mathbf{e} = 0$, by differentiation with respect to u and v we get

$$\mathbf{e}_u.\mathbf{r}_u + \mathbf{e}.\mathbf{r}_{uu} = 0, \quad \mathbf{e}_v.\mathbf{r}_u + \mathbf{e}.\mathbf{r}_{uv} = 0,$$

$$\mathbf{e}_u.\mathbf{r}_v + \mathbf{e}.\mathbf{r}_{vu} = 0, \quad \mathbf{e}_v.\mathbf{r}_v + \mathbf{e}.\mathbf{r}_{vv} = 0.$$

Using our previous values for L, M, N we now see that

$$L = -\,\mathbf{r}_u.\mathbf{e}_u, \quad M = -\,(\mathbf{r}_u.\mathbf{e}_v + \mathbf{r}_v.\mathbf{e}_u)/2, \quad N = -\,\mathbf{r}_v.\mathbf{e}_v,$$

so that

$$\kappa \cos \theta = \frac{L du^2 + 2M du dv + N dv^2}{E du^2 + 2F du dv + G dv^2}.$$

It is therefore clear that all curves Γ through P with the same tangent $\mathbf{t}$ and principal normal $\mathbf{p}$ have the same curvature κ at P, and in particular, if $\boldsymbol{\kappa}_G = 0$ for any curve Γ, its principal normal will coincide with $\mathbf{e}$, and so its curvature $|\boldsymbol{\kappa}_N|$ is the same as the curvature κ_N of the normal section with the same tangent $\mathbf{t}$.

Curvatures of the Normal Sections through a Point of the Surface

In order to examine the nature of the neighbourhood of a particular point P on S it is useful to consider the curvatures of the normal sections through P. The different sections are obtained by allowing the ratio du/dv to take all values. This means that the quadratic form $Ldu^2 + 2Mdudv + Ndv^2$ may take positive, zero or negative values. Thus we could only preserve the sign of κ_N by taking the direction of the normal to S at P sometimes in one direction and sometimes in the reverse direction. It is simpler to fix the direction of the normal to S, and allow the sign of κ_N to vary. This means that if κ_N takes opposite signs for two different normal sections through the same point, the sections lie on opposite sides of the tangent plane at P in the neighbourhood of P; and if κ_N is zero, in the simplest case the section will have an inflexional tangent at P. Now

$$\kappa_N = \frac{(Ldu^2 + 2Mdudv + Ndv^2)}{(Edu^2 + 2Fdudv + Gdv^2)}$$

is a rational quadratic function of the ratio du/dv, with a denominator essentially positive, since $(EG - F^2) > 0$ and $E > 0$, and it can only take finite values. Writing this equation as

$$du^2(\kappa_N E - L) + 2dudv(\kappa_N F - M) + dv^2(\kappa_N G - N) = 0,$$

we see that for a given value of κ_N there are two possible values for the ratio du/dv, which may coincide, except in the case where all the coefficients vanish. So there are two normal sections, which may coincide, with the same curvature, apart from the exception.

Alternatively we may regard this equation as giving κ_N as a function of du/dv, in which case the extreme values of κ_N are obtained by making the partial derivatives with regard to du

and dv of the left-hand member of the equation both vanish. This gives

$$(\kappa_N E - L)du + (\kappa_N F - M)dv = 0,$$
$$(\kappa_N F - M)du + (\kappa_N G - N)dv = 0.$$

Eliminating du, dv, the extreme values of κ_N are given by

$$\begin{vmatrix} \kappa_N E - L & \kappa_N F - M \\ \kappa_N F - M & \kappa_N G - N \end{vmatrix} = 0,$$

or

$$\kappa_N^2(EG - F^2) - \kappa_N(EN + 2FM + GL) + (LN - M^2) = 0.$$

If these extreme values of κ_N are K_1, K_2, then

$$K_1 + K_2 = (EN + 2FM + GL)/(EG - F^2),$$
$$K_1 K_2 = (LN - M^2)/(EG - F^2).$$

Definition 3.11. *K_1, K_2 are called the* **principal curvatures** *of S at P, $(K_1 + K_2)/2$ the* **mean curvature**, *and $K \equiv K_1 K_2$ the* **total** (*or* **Gauss**) **curvature** *of the surface at this point.*

The exception mentioned above, in which all the coefficients of the equation vanish, gives $\kappa_N = L/E = M/F = N/G$, and the curvature is the same for all normal sections. We give

Definition 3.12. *A point U at which all the normal sections have the same curvature is called an* **umbilic.**

Lines of Curvature

If, instead of eliminating du, dv from the equations giving the extreme values of κ_N, we eliminate κ_N, we get

$$\begin{vmatrix} Edu + Fdv & Ldu + Mdv \\ Fdu + Gdv & Mdu + Ndv \end{vmatrix} = 0,$$

which reduces to

$$\begin{vmatrix} dv^2 & -dvdu & du^2 \\ E & F & G \\ L & M & N \end{vmatrix} = 0,$$

that is

$$du^2(EM - FL) + dudv(EN - GL) + dv^2(FN - GM) = 0.$$

Regarded as a quadratic in du/dv this equation gives the direction of the tangents to the normal sections with extreme curvatures, the **principal directions**; regarding it as a differential equation we may solve it, and so obtain two families of curves on S of the form $\phi_1(u,v,\lambda) = 0$, $\phi_2(u,v,\mu) = 0$, λ and μ being parameters determining the members of the family.

One curve from each family passes through any point P of S, and we have

Definition 3.13. *A curve on S whose direction at any point P is a principal direction is called a* **line of curvature.**

Conjugate Directions

Definition 3.14. *Two directions at P given by the ratios* du_1/dv_1, du_2/dv_2 *are called* **conjugate directions** *if*

$$Ldu_1du_2 + M(du_1dv_2 + du_2dv_1) + Ndv_1dv_2 = 0.$$

Theorem 3.3. *The two lines of curvature through any point on a surface are (i) perpendicular, (ii) in conjugate directions.*

From the equation for the lines of curvature we have

$$(du_1/dv_1)(du_2/dv_2) = (FN - GM)/(EM - FL),$$

$$(du_1/dv_1) + (du_2/dv_2) = (GL - EN)/(EM - FL).$$

So, if $d\mathbf{r}_1$, $d\mathbf{r}_2$ are in the directions of the tangents

$$\begin{aligned} d\mathbf{r}_1.d\mathbf{r}_2 &= (\mathbf{r}_u\,du_1 + \mathbf{r}_v\,dv_1)\,.\,(\mathbf{r}_u\,du_2 + \mathbf{r}_v\,dv_2) \\ &= Edu_1du_2 + F(du_1dv_2 + du_2dv_1) + Gdv_1dv_2 \\ &= [E(FN - GM) + F(GL - EN) + \\ &\qquad + G(EM - FL)]dv_1dv_2/(EM - FL) \\ &= 0, \end{aligned}$$

and the lines are perpendicular. Also

$$\begin{aligned} &Ldu_1du_2 + M(du_1dv_2 + du_2dv_1) + Ndv_1dv_2 \\ &\quad = [L(FN - GM) + M(GL - EN) + \\ &\qquad\qquad + N(EM - FL)]dv_1dv_2/(EM - FL) \\ &\quad = 0, \end{aligned}$$

and the lines are conjugate.

Asymptotic Lines

At any point there are two directions which are self-conjugate given by the equation

$$Ldu^2 + 2Mdudv + Ndv^2 = 0.$$

Definition 3.15. *Lines on the surface whose directions satisfy this equation are called* **asymptotic lines,** *and the directions are known as* **asymptotic directions.**

In the case of the lines of curvature there were two, possibly coincident, lines passing through each point, both of them real. However the equation for the asymptotic lines has real roots only if $LN \leqslant M^2$, and the equations of the two families of asymptotic lines are found by solving the differential equation. Now the Gauss curvature, K, was $(LN - M^2)/(EG - F^2)$, and $(EG - F^2) > 0$. So the asymptotic lines only exist if $K \leqslant 0$, that is if K_1, K_2 are of opposite sign, or if one of them at least is zero. When they are of opposite sign we have already seen that the principal sections lie on opposite sides of the tangent plane to the surface, as, for example, in the case of the hyperboloid of one sheet. In the case when $K = 0$ at all points, the surface is of a special type called a **developable surface.** We shall deal with developable surfaces later.

Theorem 3.4. *The osculating plane at any point on an asymptotic line is the tangent plane to the surface at that point.*

If $\mathbf{e}$ is a unit normal to the surface at P and $\boldsymbol{\kappa}$ the curvature vector of the asymptotic line at this point, we have already shown that $\boldsymbol{\kappa}.\mathbf{e} + (d\mathbf{r}.d\mathbf{e})/ds^2$ is zero; and since the line is asymptotic

$$-d\mathbf{r}.d\mathbf{e} = Ldu^2 + 2Mdudv + Ndv^2 = 0.$$

Thus $\boldsymbol{\kappa}.\mathbf{e} = 0$, giving $\boldsymbol{\kappa}$ perpendicular to $\mathbf{e}$ and so equal to $\boldsymbol{\kappa}_G$. The normal curvature vector $\boldsymbol{\kappa}_N$ is therefore zero, and the principal normal $\mathbf{p}$ of the asymptotic line lies in the tangent plane to the surface. So this plane is the osculating plane.

The Geometrical Significance of the Principal and Self-conjugate Directions

Suppose at any point P on a surface S we take rectangular cartesian axes Px, Py lying in the tangent plane at P, and Pz perpendicular to this plane. Then the equation of the surface may be written

$$z = ax^2 + 2hxy + by^2 + \text{(terms involving higher powers of } x \text{ and } y).$$

If c is a small constant, then the section of S by the plane $z = c$ approximates to the conic $ax^2 + 2hxy + by^2 = c$, whose asymptotes are given by

$$ax^2 + 2hxy + by^2 = 0,$$

and whose axes are given by

$$\frac{x^2 - y^2}{a - b} = \frac{xy}{h}.$$

Now using x, y as parametric coordinates for S, the position vector $\mathbf{r}$, measured from P, of a neighbouring point Q on the surface may be written approximately as

$$\mathbf{r} = \mathbf{i}x + \mathbf{j}y + \mathbf{k}(ax^2 + 2hxy + by^2),$$

where x, y are taken to be small. So

$$\mathbf{r}_x = \mathbf{i} + 2\mathbf{k}(ax + hy), \quad \mathbf{r}_y = \mathbf{j} + 2\mathbf{k}(hx + by),$$

$$\mathbf{r}_{xx} = 2\mathbf{k}a, \quad \mathbf{r}_{xy} = 2\mathbf{k}h, \quad \mathbf{r}_{yy} = 2\mathbf{k}b.$$

This gives at the point P where x and y are zero

$$E = 1, \ F = 0, \ G = 1, \ L = 2a/W, \ M = 2h/W, \ N = 2b/W,$$

where $W = (EG - F^2)^{1/2} = 1$. So the directions of the asymptotic lines and the lines of curvature at P are given by the values of the ratio dx/dy satisfying respectively the equations

$$adx^2 + 2hdxdy + bdy^2 = 0,$$

$$\begin{vmatrix} dy^2 & -dydx & dx^2 \\ 1 & 0 & 1 \\ a & h & b \end{vmatrix} = 0,$$

and these correspond to the directions of the asymptotes and axes of the conic referred to above. We notice at once that a plane section of S taken parallel and near to the tangent plane at P is approximately elliptical, parabolic, or hyperbolic according as the asymptotic lines are non-existent, coincident, or real.

In addition, two diameters of the conic whose gradients are m_1, m_2 are conjugate when $a + h(m_1 + m_2) + bm_1m_2 = 0$, and so are parallel to a pair of conjugate directions on S given by the equation

$$Ldx_1dx_2 + M(dx_1dy_2 + dx_2dy_1) + Ndy_1dy_2 = 0.$$

Geometrical Method of Obtaining Conjugate Directions

A somewhat intuitive approach to conjugate directions is to consider the tangent planes π and σ to the surface at two neighbouring points P and Q lying on a curve Γ on S whose direction at P is given by the ratio du_1/dv_1. Then if $\mathbf{m} + \delta\mathbf{m}$ is a vector along the line of intersection of π and σ, with $\delta\mathbf{m} \to 0$ as $\sigma \to \pi$, then $\mathbf{m} + \delta\mathbf{m}$ is perpendicular to the normals to π and σ. Calling these respectively $\mathbf{e}$ and $\mathbf{e} + \delta\mathbf{e}$ we have $(\mathbf{m} + \delta\mathbf{m}).(\mathbf{e} + \delta\mathbf{e}) = 0$ and $(\mathbf{m} + \delta\mathbf{m}).\mathbf{e} = 0$. So $(\mathbf{m} + \delta\mathbf{m}).\delta\mathbf{e} = 0$. Now if δs is the length of the arc PQ, as Q approaches P we get $\mathbf{m}.(d\mathbf{e}/ds) = 0$, with $\mathbf{m}$ the tangent to S at P in a direction given by du_2/dv_2. Then the direction of $\mathbf{m}$ is conjugate to the direction of Γ at P. For if we take $d\mathbf{r}$ to be a vector along $\mathbf{m}$ we get at once $d\mathbf{r}.(d\mathbf{e}/ds) = 0$, giving

$$(\mathbf{r}_u du_2 + \mathbf{r}_v dv_2) . (\mathbf{e}_u du_1 + \mathbf{e}_v dv_1) = 0,$$

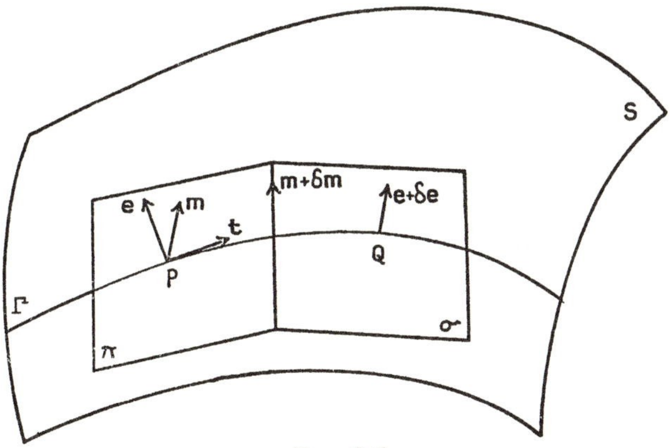

FIG. 3.6

and multiplying out this gives

$$Ldu_1du_2 + M(du_1dv_2 + du_2dv_1) + Ndv_1dv_2 = 0,$$

which is the equation for a pair of conjugate directions.

Theorems on Lines of Curvature

An interesting result due to **Dupin** is

Theorem 3.5. *If three families of surfaces are such as to cut one another orthogonally at each of their common points of intersection the curves in which they intersect must be lines of curvature on the surfaces.*

Suppose we consider the position vector $\mathbf{r}(u,v,w)$ which for any triad of values of u, v, w represents some point in three-dimensional space. We have already in our definition 3.1 defined a surface as the locus given by the extremities of the position vector $\mathbf{r}$ where $\mathbf{r}$ is a function of two real parameters u and v. So, if we make w a constant, w_0 say, the locus of the extremities of $\mathbf{r}(u,v,w_0)$ will be a surface, and by allowing w_0 to take a whole set of values we shall obtain a family of surfaces. It is then clear that by taking in turn $u =$ const., $v =$ const., $w =$ const., and allowing the constants to vary we shall arrive at three families of surfaces.

Three constants u_0, v_0, w_0 will determine three surfaces, one from each family, and each pair of surfaces will intersect in a curve, for points common to the surfaces given by v_0, w_0 will lie at the extremities of the vectors $\mathbf{r}(u,v_0,w_0)$, which being a system of one-parameter vectors define a curve, the curve of intersection of the surfaces.

At a common point of the surfaces given by u_0, v_0, w_0 the vectors $\mathbf{r}_u(u,v_0,w_0)$, $\mathbf{r}_v(u_0,v,w_0)$, $\mathbf{r}_w(u_0,v_0,w)$ are tangential to the curves of intersection of the surfaces taken in pairs, and since it is given that these curves are orthogonal in pairs, at such common points we have

$$\mathbf{r}_v.\mathbf{r}_w = 0, \quad \mathbf{r}_w.\mathbf{r}_u = 0, \quad \mathbf{r}_u.\mathbf{r}_v = 0.$$

So differentiating we get

$$\mathbf{r}_{vu}.\mathbf{r}_w + \mathbf{r}_v.\mathbf{r}_{wu} = 0, \quad \mathbf{r}_{wv}.\mathbf{r}_u + \mathbf{r}_w.\mathbf{r}_{uv} = 0,$$
$$\mathbf{r}_{uw}.\mathbf{r}_v + \mathbf{r}_u.\mathbf{r}_{vw} = 0,$$

and therefore

$$2(\mathbf{r}_u.\mathbf{r}_{vw} + \mathbf{r}_v.\mathbf{r}_{wu} + \mathbf{r}_w.\mathbf{r}_{uv}) = 0,$$

which gives

$$\mathbf{r}_u.\mathbf{r}_{vw} = 0.$$

Therefore $\mathbf{r}_v$, $\mathbf{r}_w$, $\mathbf{r}_{vw}$ are all perpendicular to $\mathbf{r}_u$, so

$$[\mathbf{r}_v, \mathbf{r}_w, \mathbf{r}_{vw}] = 0,$$

and on the surface $u = u_0$ we get $M = 0$. Also, since $\mathbf{r}_v.\mathbf{r}_w = 0$, we have $F = 0$. So the lines of curvature on this surface are

$$\begin{vmatrix} dw^2 & -dwdv & dv^2 \\ E & 0 & G \\ L & 0 & N \end{vmatrix} = 0,$$

that is $dwdv = 0$, or $v =$ const., $w =$ const., and similarly on the other two surfaces.

An important theorem due to **Olinde Rodrigues** is

Theorem 3.6. *At any point P on a line of curvature the differentials of the position vector of P and the normal to the surface satisfy the relation $\kappa d\mathbf{r} + d\mathbf{e} = \mathbf{0}$, where κ is the curvature of the curve at P.*

Along a line of curvature (see page 52) we have

$$Ldu + Mdv - \kappa(Edu + Fdv) = 0,$$

$$Mdu + Ndv - \kappa(Fdu + Gdv) = 0,$$

and inserting the values for the coefficients of du and dv we get

$$-\mathbf{r}_u.\mathbf{e}_u du - \mathbf{r}_u.\mathbf{e}_v dv - \kappa(\mathbf{r}_u.\mathbf{r}_u du + \mathbf{r}_u.\mathbf{r}_v dv) = 0,$$

$$-\mathbf{r}_v.\mathbf{e}_u du - \mathbf{r}_v.\mathbf{e}_v dv - \kappa(\mathbf{r}_v.\mathbf{r}_u du + \mathbf{r}_v.\mathbf{r}_v dv) = 0,$$

that is

$$\mathbf{r}_u.(d\mathbf{e} + \kappa d\mathbf{r}) = 0, \quad \mathbf{r}_v.(d\mathbf{e} + \kappa d\mathbf{r}) = 0.$$

Now as $\mathbf{e}.\mathbf{e} = 1$, we have $\mathbf{e}.d\mathbf{e} = 0$, and since $\mathbf{e}$ is perpendicular to $d\mathbf{r}$, we get also $\mathbf{e}.d\mathbf{r} = 0$. There is therefore a third equation $\mathbf{e}.(d\mathbf{e} + \kappa d\mathbf{r}) = 0$ which, taken together with the two equations above, shows that

$$d\mathbf{e} + \kappa d\mathbf{r} = \mathbf{0},$$

since $\mathbf{r}_u$, $\mathbf{r}_v$, and $\mathbf{e}$ are linearly independent vectors.

The converse theorem also holds, for, given this relation, the scalar products of the left member with $\mathbf{r}_u$ and $\mathbf{r}_v$ lead at once to the two equations above, and so to the equations for the lines of curvature.

An immediate deduction from this relation is

Theorem 3.7. *Along a line of curvature*

$$\lim_{\delta s \to 0} [\delta\mathbf{r}, \mathbf{e}, \mathbf{e} + \delta\mathbf{e}]/\delta s^2 = 0,$$

for this limit is

$$\lim_{\delta s \to 0} [\delta\mathbf{r}/\delta s, \mathbf{e}, \delta\mathbf{e}/\delta s] = [d\mathbf{r}/ds, \mathbf{e}, d\mathbf{e}/ds],$$

which is zero since $d\mathbf{e}/ds = -\kappa d\mathbf{r}/ds$.

Now this relation may be interpreted as stating that 'consecutive' normals to a surface along a line of curvature 'intersect'. So the surface formed by them may be developed into a plane, a fact which gives (*see pages* 65, 66, 67)

Theorem 3.8. *Normals to a surface along a line of curvature lie on a developable surface.*

Another result, due to **Joachimsthal,** is

Theorem 3.9. *If two of the following statements hold for a pair of intersecting surfaces S_1, S_2, then so does the third.*

1. *The curve of intersection of S_1, S_2 is a line of curvature on S_1,*
2. *The curve of intersection of S_1, S_2 is a line of curvature on S_2,*
3. *S_1 and S_2 intersect at a constant angle.*

If θ is the angle between the surfaces at any point on their curve of intersection then $\cos\theta = \mathbf{e}_1.\mathbf{e}_2$, where $\mathbf{e}_1$, $\mathbf{e}_2$ are the normals to the surfaces at the point. So

$$\frac{d}{ds}(\cos\theta) = \frac{d\mathbf{e}_1}{ds}\cdot\mathbf{e}_2 + \frac{d\mathbf{e}_2}{ds}\cdot\mathbf{e}_1,$$

where s is the arc-length of the curve.

If the curve is a line of curvature on S_1 then $d\mathbf{e}_1/ds = -\kappa_1\mathbf{t}$, $\mathbf{t}$ being the tangent to the curve and κ_1 the normal curvature for S_1. But $\mathbf{e}_1.\mathbf{t} = 0$ and $\mathbf{e}_2.\mathbf{t} = 0$, so

$$\frac{d}{ds}(\cos\theta) = \frac{d\mathbf{e}_2}{ds}\cdot\mathbf{e}_1,$$

which will be zero if the curve is also a line of curvature on S_2, so giving $\cos\theta$ to be a constant.

If, on the other hand, it is given that $\cos\theta$ is constant, and that the curve is a line of curvature on S_1, then $\mathbf{e}_1.(d\mathbf{e}_2/ds) = 0$, and so $d\mathbf{e}_2/ds$ is perpendicular to both $\mathbf{e}_1$ and $\mathbf{e}_2$, and is therefore parallel to $\mathbf{t}$, which shows that the curve is also a line of curvature on S_2, and writing $d\mathbf{e}_2/ds = -\kappa_2\mathbf{t}$ makes κ_2 the normal curvature for S_2.

The Gauss (or Total) Curvature of a Surface

The curvature of a space curve is the rate of turn of the tangent per arc. It may be obtained by considering the curve traced out by the extremity of a unit vector drawn from some origin O to be parallel to the tangent to the curve as the tangent moves along the curve. This new curve lies on a sphere of radius unity, centre O, and the limiting value of the ratio of the length of an arc of it to the length of the corresponding arc on the original space curve, as the length of the arc approaches zero, is evidently the curvature $\kappa = d\psi/ds$ of the space curve, for the angle $\delta\psi$ between two neighbouring tangents to the space curve is also the angle between the corresponding vectors drawn from O, and, since the radius of the sphere is unity, $\delta\psi$ is also approximately the length of an element of arc of the curve traced out on the sphere. So the ratio

$$\text{limit}\left[\frac{\text{arc-length on sphere}}{\text{arc-length on space curve}}\right] = \frac{d\psi}{ds} = \kappa.$$

Similarly, if we draw unit vectors from O to be parallel to the normals to a surface S at the points of the boundary of an area δS on S, their extremities will trace out the boundary of an area $\delta\Omega$ on the surface of the unit sphere. A natural measure of the amount of bending of S at a point P would be given by taking the limit of the ratio $\delta\Omega/\delta S$ as δS shrinks to zero at P, assuming this to be independent of the area chosen on S and the manner in which it is shrunk to zero. Using for our areas the elements of area as previously defined on page 46, we prove

Theorem 3.10. *The limit of the ratio $\delta\Omega/\delta S$ is the total curvature of the surface at P.*

Take the lines of curvature on S as parametric lines. The element of area is

$$dS = |(\mathbf{r}_u du \wedge \mathbf{r}_v dv).\mathbf{e}| = |[\mathbf{r}_u, \mathbf{r}_v, \mathbf{e}]dudv|.$$

The corresponding element of area for the unit sphere is

$$d\Omega = |(\mathbf{e}_u du \wedge \mathbf{e}_v dv).\mathbf{e}| = |[\mathbf{e}_u, \mathbf{e}_v, \mathbf{e}]dudv|,$$

for the two surfaces have their unit normals parallel at the corresponding points. By the theorem of Olinde Rodrigues for a line of curvature,

$$\mathbf{e}_u + \kappa_1 \mathbf{r}_u = \mathbf{0}, \quad \mathbf{e}_v + \kappa_2 \mathbf{r}_v = \mathbf{0},$$

so

$$\lim_{\delta S \to 0} (\delta\Omega/\delta S) = d\Omega/dS = \kappa_1 \kappa_2 = K,$$

the total curvature of S at P.

Distortion of Surfaces

If it is possible to bend one surface into another without in any way stretching it, then it is clear that the lengths of corresponding curves on the two surfaces will be the same. In fact if we have covered the one surface with a system of parametric curves the corresponding curves on the other surface will form a parametric system with the same two parameters, and the line-element for the two surfaces, whose square is called the metric of the surfaces, will remain the same. We therefore give

Definition 3.16. *Two surfaces are called applicable, or isometric, if they have the same metric* $ds^2 = Edu^2 + 2Fdudv + Gdv^2$.

It may not always be possible to carry out the distortion without making cuts in the surface. For example a frustum of an open circular cone, such as a lampshade, may be distorted into a portion of a cone which is no longer circular, but to distort it into a plane it would be sufficient to make a cut along one of the generators. Again, a circular cylinder can be deformed into an elliptic cylinder, or a cylinder with any sort of closed curve for its section, but it can only be made into a plane if we cut it along one of the generators. If, however, we consider the surface of a sphere, although it can be distorted by making dents in it, leaving the metric unaltered, it is impossible to distort it even if we make cuts in it, in such a manner that it can be made into a plane.

We now prove a theorem, due to **Gauss**, which shows that the total curvature at any point on a surface depends only on the metric of the surface, and this shows that isometric surfaces have the same total curvature at corresponding points. In particular, any surface which is isometric to a plane will have zero total curvature, and this property, $K = 0$, is characteristic of the particular class of surfaces known as **developable surfaces**, a fact which will be proved later on pages 72, 73.

Theorem 3.11. (*The* **'Theorema Egregium'** *of* **Gauss.**) *The total curvature K at any point P of a surface depends only on the values at P of E, F, G and their derivatives.*

It has already been shown that $K = (LN - M^2)/(EG - F^2)$ and so, with the values of E, F, G, L, M, N previously obtained and with $W = (EG - F^2)^{1/2}$, we consider the expression

$$KW^4 = [\mathbf{r}_u,\mathbf{r}_v,\mathbf{r}_{uu}][\mathbf{r}_u,\mathbf{r}_v,\mathbf{r}_{vv}] - [\mathbf{r}_u,\mathbf{r}_v,\mathbf{r}_{uv}]^2.$$

By regarding the vectors $\mathbf{r}_u$, $\mathbf{r}_v$, $\mathbf{r}_{uu}$, . . . as column vectors, and applying the multiplication theorem for determinants in the column by column form, we have

$$KW^4 = \begin{vmatrix} \mathbf{r}_u\cdot\mathbf{r}_u & \mathbf{r}_u\cdot\mathbf{r}_v & \mathbf{r}_u\cdot\mathbf{r}_{vv} \\ \mathbf{r}_v\cdot\mathbf{r}_u & \mathbf{r}_v\cdot\mathbf{r}_v & \mathbf{r}_v\cdot\mathbf{r}_{vv} \\ \mathbf{r}_{uu}\cdot\mathbf{r}_u & \mathbf{r}_{uu}\cdot\mathbf{r}_v & \mathbf{r}_{uu}\cdot\mathbf{r}_{vv} \end{vmatrix} - \begin{vmatrix} \mathbf{r}_u\cdot\mathbf{r}_u & \mathbf{r}_u\cdot\mathbf{r}_v & \mathbf{r}_u\cdot\mathbf{r}_{uv} \\ \mathbf{r}_v\cdot\mathbf{r}_u & \mathbf{r}_v\cdot\mathbf{r}_v & \mathbf{r}_v\cdot\mathbf{r}_{uv} \\ \mathbf{r}_{uv}\cdot\mathbf{r}_u & \mathbf{r}_{uv}\cdot\mathbf{r}_v & \mathbf{r}_{uv}\cdot\mathbf{r}_{uv} \end{vmatrix}$$

$$= \begin{vmatrix} E & F & \mathbf{r}_u\cdot\mathbf{r}_{vv} \\ F & G & \mathbf{r}_v\cdot\mathbf{r}_{vv} \\ \mathbf{r}_{uu}\cdot\mathbf{r}_u & \mathbf{r}_{uu}\cdot\mathbf{r}_v & \mathbf{r}_{uu}\cdot\mathbf{r}_{vv} \end{vmatrix} - \begin{vmatrix} E & F & \mathbf{r}_u\cdot\mathbf{r}_{uv} \\ F & G & \mathbf{r}_v\cdot\mathbf{r}_{uv} \\ \mathbf{r}_{uv}\cdot\mathbf{r}_u & \mathbf{r}_{uv}\cdot\mathbf{r}_v & \mathbf{r}_{uv}\cdot\mathbf{r}_{uv} \end{vmatrix}$$

$$= \begin{vmatrix} E & F & \mathbf{r}_u\cdot\mathbf{r}_{vv} \\ F & G & \mathbf{r}_v\cdot\mathbf{r}_{vv} \\ \mathbf{r}_{uu}\cdot\mathbf{r}_u & \mathbf{r}_{uu}\cdot\mathbf{r}_v & (\mathbf{r}_{uu}\cdot\mathbf{r}_{vv}-\mathbf{r}_{uv}^2) \end{vmatrix} - \begin{vmatrix} E & F & \mathbf{r}_u\cdot\mathbf{r}_{uv} \\ F & G & \mathbf{r}_v\cdot\mathbf{r}_{uv} \\ \mathbf{r}_{uv}\cdot\mathbf{r}_u & \mathbf{r}_{uv}\cdot\mathbf{r}_v & 0 \end{vmatrix}$$

for by expanding the determinants using their last columns we see that the term $\mathbf{r}_{uv}.\mathbf{r}_{uv}(EG - F^2)$ has merely been transferred from the second determinant to the first.

But from $E = \mathbf{r}_u.\mathbf{r}_u$, $F = \mathbf{r}_u.\mathbf{r}_v$, $G = \mathbf{r}_v.\mathbf{r}_v$ we get

$$\begin{aligned} E_u/2 &= \mathbf{r}_u.\mathbf{r}_{uu}, \quad F_u = \mathbf{r}_{uu}.\mathbf{r}_v + \mathbf{r}_u.\mathbf{r}_{uv}, \quad G_u/2 = \mathbf{r}_v.\mathbf{r}_{uv}, \\ E_v/2 &= \mathbf{r}_u.\mathbf{r}_{uv}, \quad F_v = \mathbf{r}_{uv}.\mathbf{r}_v + \mathbf{r}_u.\mathbf{r}_{vv}, \quad G_v/2 = \mathbf{r}_v.\mathbf{r}_{vv}, \\ E_{vv}/2 &= \mathbf{r}_{uv}.\mathbf{r}_{uv} + \mathbf{r}_u.\mathbf{r}_{uvv}, \quad G_{uu}/2 = \mathbf{r}_{vu}.\mathbf{r}_{uv} + \mathbf{r}_v.\mathbf{r}_{uvu}, \\ F_{uv} &= \mathbf{r}_{uuv}.\mathbf{r}_v + \mathbf{r}_{uu}.\mathbf{r}_{vv} + \mathbf{r}_{uv}.\mathbf{r}_{uv} + \mathbf{r}_u.\mathbf{r}_{uvv}, \end{aligned}$$

so

$$F_u - E_v/2 = \mathbf{r}_{uu}.\mathbf{r}_v, \quad F_v - G_u/2 = \mathbf{r}_{vv}.\mathbf{r}_u,$$

and

$$\begin{aligned} E_{vv}/2 + G_{uu}/2 &= 2\mathbf{r}_{uv}.\mathbf{r}_{uv} + \mathbf{r}_u.\mathbf{r}_{uvv} + \\ &\quad + \mathbf{r}_v.\mathbf{r}_{vuu} = 2\mathbf{r}_{uv}^2 + F_{uv} - \mathbf{r}_{uu}.\mathbf{r}_{vv} - \mathbf{r}_{uv}^2. \end{aligned}$$

Therefore

$$\mathbf{r}_{uu}.\mathbf{r}_{vv} - \mathbf{r}_{uv}^2 = -E_{vv}/2 + F_{uv} - G_{uu}/2.$$

This gives $KW^4 =$

$$\begin{vmatrix} E & F & F_v - G_u/2 \\ F & G & G_v/2 \\ E_u/2 & F_u - E_v/2 & -G_{uu}/2 + F_{uv} - E_{vv}/2 \end{vmatrix} - \begin{vmatrix} E & F & E_v/2 \\ F & G & G_u/2 \\ E_v/2 & G_u/2 & 0 \end{vmatrix},$$

showing that K depends only on E, F, G and their derivatives.

Developable Surfaces

We begin with an informal geometrical discussion.

Suppose the equation of a plane is dependent on some parameter, and that as the parameter varies continuously the equation varies continuously. We then get a family of planes which may be regarded as being generated by the continuous motion of a plane. Any two 'consecutive' planes will meet in a line. By picking out a whole series of 'consecutive' planes, and turning each about the line of intersection with its neighbour, until it lies in the plane of its neighbour, we can flatten the whole series of planes into one single plane. If this process is made continuous, each line of intersection of a 'consecutive' pair of planes will have a limiting position, and the whole set of such lines in their limiting positions will form a ruled surface, in this case a surface which can be formed by the continuous motion of a line which depends only on one parameter, two 'consecutive' lines being such that they always intersect. The ruled surface so generated can thus be flattened into a single plane. In the same way, if we consider three 'consecutive' planes they will possess a pair of lines of intersection lying in the intermediate plane, and this pair of lines will cut in a point. Again, making the process continuous, we see that the ruled surface consists of a set of lines, and that each line contains a special single point, the limit of the intersection of the line with its neighbour. The entire set of these points will lie on a curve, which itself lies on the ruled surface. By analogy with the envelope of a one-parameter family of lines in a plane, we see that the ruled surface is the envelope of the family of planes, and that any portion of such a surface may be folded, without stretching or tearing, into a portion of a single plane. Notice that we have talked somewhat generally, assuming that our planes do intersect in lines and that our lines do intersect in points. Cases in which these properties do not hold require special examination.

As the vector equation of a plane may be written $(\mathbf{r} - \mathbf{n}).\mathbf{n} = 0$, where $\mathbf{r}$ is the position vector to any point on the plane, and $\mathbf{n}$ is the vector to the foot of the perpendicular from the origin to the plane, if $\mathbf{n}$ is a function of some parameter t, we may obtain the envelope of the family of planes obtained by varying t by eliminating t between the equations

$(\mathbf{r}-\mathbf{n}).\mathbf{n}=0$ and $(\mathbf{r}-\mathbf{n}).\dot{\mathbf{n}}=\mathbf{n}.\dot{\mathbf{n}}$, that is to say by getting the locus of the limiting positions of the lines of intersection of two neighbouring planes of the family. This locus is, of course, a ruled surface, being a locus constructed from the motion of a line. We give as a definition

Definition 3.17. *The envelope of a one-parameter family of planes is called a* **developable surface.**

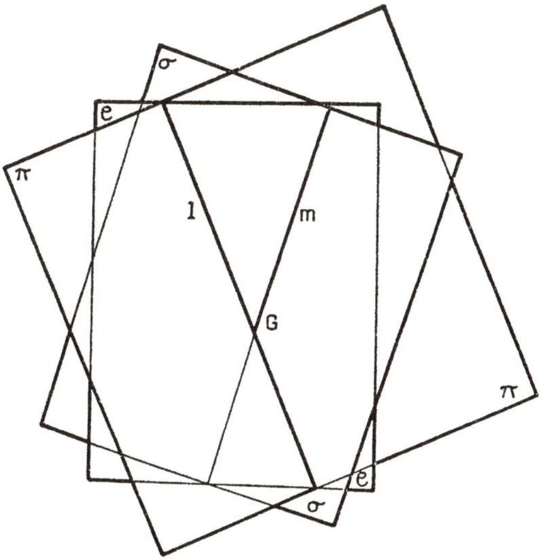

FIG. 3.7

Three consecutive planes π, ρ, σ intersecting in consecutive lines l, m which intersect at G. As π, ρ, σ approach coincidence, l, m approach generators lying on the surface Σ enveloped by the system of planes, and G approaches a point on the edge of regression, a curve lying on the surface Σ.

It should at once be noted that not all ruled surfaces are developable; an example which immediately occurs to mind is the hyperboloid of one sheet whose equation is $x^2/a^2 - y^2/b^2 + z^2/c^2 = 1$. This surface contains two sets of lines, but is certainly not developable.

We also give as a definition

Definition 3.18. *The set of lines on a developable surface is called a set of* **generators,** *or* **characteristic lines,** *and the curve formed by the special points obtained as the limits of the intersection of two neighbouring generators is known as the* **edge of regression** *of the surface.*

These special points are given as the intersections of the three planes $\mathbf{r}.\mathbf{n} = c$, $\mathbf{r}.\dot{\mathbf{n}} = \dot{c}$, $\mathbf{r}.\ddot{\mathbf{n}} = \ddot{c}$, where $c = \mathbf{n}.\mathbf{n}$.

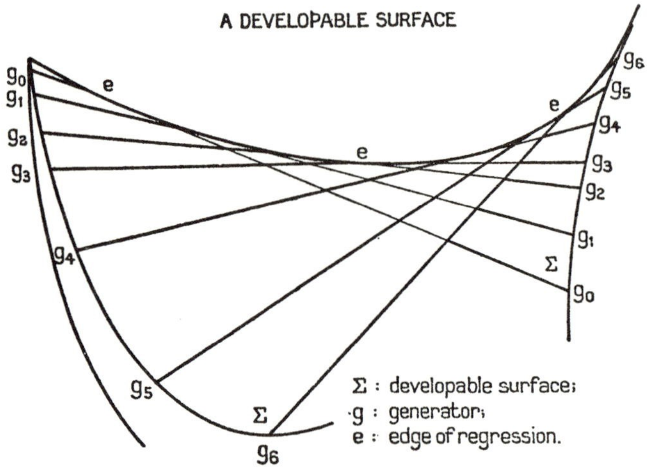

FIG. 3.8

The two simplest developable surfaces are the cylinder and the cone. In the first, the generators are all parallel, and the edge of regression reduces to the single 'point at infinity' where the generators may be said to meet; in the second, all generators meet at the vertex, and so the edge of regression reduces to a single point, the vertex itself. To develop these surfaces into planes we first need to make a cut along one of the generators in each case.

Analytically, we may prove that

Theorem 3.12. *Any developable surface is the locus generated by the tangent lines to a skew curve.*

This is intuitively apparent from the geometrical point of view; for a developable surface Σ is the locus of the characteristic lines of a one-parameter family of planes, and these lines are all tangents to the edge of regression.

We may prove the result rigorously in the following way: let the one-parameter family of planes be

$$(\mathbf{r} - \mathbf{n}).\mathbf{n} = 0,$$

or $\mathbf{r}.\mathbf{n} = \mathbf{n}.\mathbf{n} = c$, say, where $\mathbf{r}$ is the position vector of a point on one of the planes π, whose parameter is t, and where $\mathbf{n}$, a function of t, is the normal from the origin to π.

Differentiating with respect to t we get $\mathbf{r}.\dot{\mathbf{n}} = \dot{c}$, which is a plane ρ through the characteristic line of π, the limit of the intersection of π with a neighbouring plane.

A second differentiation gives $\mathbf{r}.\ddot{\mathbf{n}} = \ddot{c}$, which is a plane σ through the limit of the common intersection of three neighbouring planes.

So the first two equations together give the generators of the ruled surface Σ, while all three equations taken together give the edge of regression of the original family of planes.

Now assuming that $\mathbf{n}$, $\dot{\mathbf{n}}$, $\ddot{\mathbf{n}}$ are three linearly independent vectors, so are $\dot{\mathbf{n}} \wedge \ddot{\mathbf{n}}$, $\ddot{\mathbf{n}} \wedge \mathbf{n}$, $\mathbf{n} \wedge \dot{\mathbf{n}}$, and we may write

$$\mathbf{r} = \lambda(\dot{\mathbf{n}} \wedge \ddot{\mathbf{n}}) + \mu(\ddot{\mathbf{n}} \wedge \mathbf{n}) + \nu(\mathbf{n} \wedge \dot{\mathbf{n}});$$

and then

$$\mathbf{r}.\mathbf{n} = \lambda[\mathbf{n},\dot{\mathbf{n}},\ddot{\mathbf{n}}] = c, \quad \mathbf{r}.\dot{\mathbf{n}} = \mu[\mathbf{n},\dot{\mathbf{n}},\ddot{\mathbf{n}}] = \dot{c}, \quad \mathbf{r}.\ddot{\mathbf{n}} = \nu[\mathbf{n},\dot{\mathbf{n}},\ddot{\mathbf{n}}] = \ddot{c}.$$

The edge of regression is thus the curve Γ given by

$$\mathbf{r} = \frac{c(\dot{\mathbf{n}} \wedge \ddot{\mathbf{n}}) + \dot{c}(\ddot{\mathbf{n}} \wedge \mathbf{n}) + \ddot{c}(\mathbf{n} \wedge \dot{\mathbf{n}})}{[\mathbf{n},\dot{\mathbf{n}},\ddot{\mathbf{n}}]}.$$

Now the tangent to Γ at t is $\dot{\mathbf{r}}$, where $\mathbf{r}$ satisfies the equations

$$\mathbf{r}.\mathbf{n} = c, \quad \mathbf{r}.\dot{\mathbf{n}} = \dot{c}, \quad \mathbf{r}.\ddot{\mathbf{n}} = \ddot{c},$$

and from the first of these $\dot{\mathbf{r}}.\mathbf{n} + \mathbf{r}.\dot{\mathbf{n}} = \dot{c}$, which, with the second, gives $\dot{\mathbf{r}}.\mathbf{n} = 0$. Also $\dot{\mathbf{r}}.\dot{\mathbf{n}} + \mathbf{r}.\ddot{\mathbf{n}} = \ddot{c}$, which, with the third, gives $\dot{\mathbf{r}}.\dot{\mathbf{n}} = 0$.

Thus $\dot{\mathbf{r}}$ is perpendicular to both $\mathbf{n}$ and $\dot{\mathbf{n}}$, the normals to π and ρ, and so lies along the line of intersection of these planes. Hence, the tangents to Γ are the generators of the developable surface Σ, and from $\dot{\mathbf{r}}.\mathbf{n} = 0$ we get $\mathbf{t}.\mathbf{n} = 0$, where $\mathbf{t}$ is the tangent to Γ, and, from the last equation, $\kappa\mathbf{p}\dot{s}.\mathbf{n} + \mathbf{t}.\dot{\mathbf{n}} = 0$, but $\mathbf{t}.\dot{\mathbf{n}} = 0$, so that both $\mathbf{t}.\mathbf{n}$ and $\mathbf{p}.\mathbf{n}$ are zero. The normal to π is therefore parallel to $\mathbf{b}$, the binormal of Γ, and the one-parameter family of planes is therefore the family of osculating planes of Γ. So we have

Theorem 3.13. *The osculating planes of the edge of regression of a developable surface form the one-parameter family of planes whose envelope is the developable surface itself.*

Standard Form of Equation for a Developable Surface

We may now write the position vector $\mathbf{R}$ of a general point Q on a developable surface Σ in the form

$$\mathbf{R} = \mathbf{r}(t) + u\dot{\mathbf{r}}(t),$$

the parameters for Σ being t and u, $\mathbf{r}(t)$ being the vector to a point P on Γ, the edge of regression, and $\dot{\mathbf{r}}(t)$ the tangent to Γ at this point.

The normal to Σ at $Q(t,u)$ is parallel to $\mathbf{R}_t \wedge \mathbf{R}_u$, that is to

$$(\dot{\mathbf{r}} + u\ddot{\mathbf{r}}) \wedge \dot{\mathbf{r}} = u\ddot{\mathbf{r}} \wedge \dot{\mathbf{r}}.$$

But the equation for the osculating plane of Γ at P is $[(\mathbf{R} - \mathbf{r}),\dot{\mathbf{r}},\ddot{\mathbf{r}}] = 0$, showing that the normal to Σ at Q is parallel to the binormal to Γ at P.

Theorem 3.14. *At all points on a generator of a developable surface the tangent plane is the same.*

For the normal to Σ at Q is parallel to $\dot{\mathbf{r}} \wedge \ddot{\mathbf{r}}$, which is independent of u.

Theorem 3.15. *The normal plane at any point on the edge of regression cuts the developable surface in a cusp.*

Take the edge of regression Γ as $\mathbf{r}(s)$, where s is the arc-length of Γ. Then Σ is $\mathbf{R} = \mathbf{r}(s) + u\mathbf{r}'(s) = \mathbf{r} + u\mathbf{t}$, where $\mathbf{t}$ is the unit tangent to Γ. We can choose s so that $s = 0$ at some arbitrary point P_0 on Γ. Then

$$\mathbf{R} = \mathbf{r}_0 + s\mathbf{t}_0 + s^2\kappa\mathbf{p}_0/2! + s^3[\kappa'\mathbf{p}_0 + \kappa(-\kappa\mathbf{t}_0 + \lambda\mathbf{b}_0)]/3! + \ldots$$
$$+ u[\mathbf{t}_0 + s\kappa\mathbf{p}_0 + s^2\{\kappa'\mathbf{p}_0 + \kappa(-\kappa\mathbf{t}_0 + \lambda\mathbf{b}_0)\}/2! + \ldots].$$

Now for $\mathbf{R}$ to lie in the normal plane to Γ at P_0, $\mathbf{R}$ must be independent of $\mathbf{t}_0$. So

$$s + s^3(-\kappa^2)/6 + \ldots + u[1 - s^2\kappa^2/2 + \ldots] = 0,$$

which gives

$$u = -[s + s^3\kappa^2/3 + \ldots],$$

making

$$\mathbf{R} = \mathbf{r}_0 + \mathbf{p}_0[s^2\kappa/2 + s^3\kappa'/6 - (s + \ldots)(\kappa s + \kappa' s^2/2 + \ldots)]$$
$$+ \mathbf{b}_0[s^3\kappa\lambda/6 + \ldots - (s + \ldots)s^2\kappa\lambda/2]$$
$$= \mathbf{r}_0 + \mathbf{p}_0[-s^2\kappa/2 + \ldots] + \mathbf{b}_0[-s^3\kappa\lambda/3 + \ldots],$$

showing that, as the coefficients of $\mathbf{p}_0$, $\mathbf{b}_0$ are of the order of s^2 and s^3 respectively, the section of the normal plane is a cusp, since the equation of the section is approximately of the form $z^2 = cy^3$, where c is some function of κ and λ.

The Ruled Surface of Tangents to a Curve Γ

We have now established the fact that any developable surface may be regarded as being generated by the tangents to a skew curve, and so may be given by an equation of the form $\mathbf{R} = \mathbf{r} + u\mathbf{t}$. From this equation, using s, the arc-length, as parameter,

$$\begin{aligned}
E &= \mathbf{R}_s.\mathbf{R}_s = (\mathbf{t} + u\kappa\mathbf{p}).(\mathbf{t} + u\kappa\mathbf{p}) = 1 + u^2\kappa^2, \\
F &= \mathbf{R}_s.\mathbf{R}_u = (\mathbf{t} + u\kappa\mathbf{p}).\mathbf{t} = 1, \\
G &= \mathbf{R}_u.\mathbf{R}_u = \mathbf{t}.\mathbf{t} = 1, \\
WL &= [\mathbf{R}_s,\mathbf{R}_u,\mathbf{R}_{ss}] = [\mathbf{t} + u\kappa\mathbf{p},\mathbf{t},\kappa\mathbf{p} + \\
&\qquad + u\kappa'\mathbf{p} + u\kappa(-\kappa\mathbf{t} + \lambda\mathbf{b})] = -u^2\kappa^2\lambda, \\
WM &= [\mathbf{R}_s,\mathbf{R}_u,\mathbf{R}_{su}] = [\mathbf{t} + u\kappa\mathbf{p},\mathbf{t},\kappa\mathbf{p}] = 0, \\
WN &= [\mathbf{R}_s,\mathbf{R}_u,\mathbf{R}_{uu}] = [\mathbf{t} + u\kappa\mathbf{p},\mathbf{t},0] = 0,
\end{aligned}$$

where $W = (EG - F^2)^{1/2} = |u\kappa|$. Now the total curvature K for any surface is $(LN - M^2)/(EG - F^2)$, and in this case $(LN - M^2)$ is zero. So we have

Theorem 3.16. *For any developable surface the total curvature is zero.*

It remains to show the converse theorem that, if the total curvature is zero, then the surface is developable. In order to do this we recollect that $L = -\mathbf{r}_u.\mathbf{e}_u$, $2M = -(\mathbf{r}_u.\mathbf{e}_v + \mathbf{r}_v.\mathbf{e}_u)$, $N = -\mathbf{r}_v.\mathbf{e}_v$, and $\mathbf{r}_u.\mathbf{e}_v = \mathbf{r}_v.\mathbf{e}_u$. Then, if $(LN - M^2)$ is zero,

$$\begin{aligned}
0 &= (\mathbf{r}_u.\mathbf{e}_u)(\mathbf{r}_v.\mathbf{e}_v) - (\mathbf{r}_u.\mathbf{e}_v)(\mathbf{r}_v.\mathbf{e}_u) \\
&= \mathbf{r}_u.[\mathbf{e}_u(\mathbf{r}_v.\mathbf{e}_v) - \mathbf{e}_v(\mathbf{r}_v.\mathbf{e}_u)] \\
&= \mathbf{r}_u.[\mathbf{r}_v \wedge (\mathbf{e}_u \wedge \mathbf{e}_v)] \\
&= (\mathbf{r}_u \wedge \mathbf{r}_v).(\mathbf{e}_u \wedge \mathbf{e}_v) \\
&= W[\mathbf{e},\mathbf{e}_u,\mathbf{e}_v].
\end{aligned}$$

Now $\mathbf{e}$ is not zero, and so this implies

either (i) $\mathbf{e}_u$ or $\mathbf{e}_v = \mathbf{0}$,

or (ii) $\mathbf{e}_u$ is parallel to $\mathbf{e}_v$,

or (iii) $\mathbf{e}$, $\mathbf{e}_u$, $\mathbf{e}_v$ are linearly dependent.

In the first case $\mathbf{e}$ is a function of one of the parameters alone. This means that the tangent planes to the surface form a one-parameter family of planes, and so the surface itself is the envelope of such a family, and therefore developable.

In the second case $\mathbf{e}_u = k(u,v)\mathbf{e}_v$, where $k(u,v)$ is some scalar function of u and v. Now make the change of variables $u_1 = u$, $v_1 = f(u,v)$, where $f(x,y) = \text{const.}$, is a solution of the differential equation $dy/dx = -k(x,y)$. Then

$$du_1 = du, \quad dv_1 = f_u du + f_v dv,$$

so

$$du = du_1, \quad f_v dv = dv_1 - f_u du.$$

This gives

$$\frac{\partial u}{\partial u_1} = 1, \quad \frac{\partial v}{\partial u_1} = -(f_u/f_v).$$

So

$$\mathbf{e}_{u_1} = \mathbf{e}_u \frac{\partial u}{\partial u_1} + \mathbf{e}_v \frac{\partial v}{\partial u_1} = \mathbf{e}_u + \mathbf{e}_v(-f_u/f_v).$$

But as $f(x,y) = \text{const.}$, $f_x + f_y(dy/dx) = 0$, that is $f_x + f_y[-k(x,y)] = 0$, and therefore

$$\mathbf{e}_{u_1} = \mathbf{e}_u + \mathbf{e}_v[-k(u,v)] = \mathbf{0}.$$

Thus with this change of variables the normal $\mathbf{e}$ depends only on the parameter v_1, and again the tangent planes are a one-parameter family, and the surface is developable.

In the third case we should have

$$\mathbf{e} = \lambda(u,v)\mathbf{e}_u + \mu(u,v)\mathbf{e}_v,$$

where $\lambda(u,v)$ and $\mu(u,v)$ are scalars. But $\mathbf{e}$ is perpendicular to $\mathbf{e}_u$ and also perpendicular to $\mathbf{e}_v$, so this is not possible.

Other conditions for ruled surfaces to be developable are given by the next two theorems.

Theorem 3.17. *For a ruled surface to be developable the tangent planes at all points of any generator must be coincident.*

Let $\mathbf{R} = \mathbf{r}_1(t) + u\mathbf{r}_2(t)$, $\mathbf{r}_2(t) \neq 0$, be the position vector of the point $P(t,u)$ on the ruled surface S. The normal to S at P is parallel to $\mathbf{R}_t \wedge \mathbf{R}_u$, that is to $(\dot{\mathbf{r}}_1 + u\dot{\mathbf{r}}_2) \wedge \mathbf{r}_2$. Along any given generator t is constant and u varies. So if the tangent plane is to remain the same for all values of u the vector product above must be a vector whose direction is independent of u. So,

$$\text{either} \quad \text{(i) } \dot{\mathbf{r}}_2 \wedge \mathbf{r}_2 = \mathbf{0}, \quad \text{or} \quad \text{(ii) } \dot{\mathbf{r}}_1 \wedge \mathbf{r}_2 = \mathbf{0}.$$

Now (i) gives $\dot{\mathbf{r}}_2 = \mathbf{0}$, or $\dot{\mathbf{r}}_2$ parallel to $\mathbf{r}_2$. In either case $\mathbf{r}_2$ is constant in direction, and so we have a cylinder, which is developable.

And (ii) gives $\dot{\mathbf{r}}_1 = \mathbf{0}$, or $\dot{\mathbf{r}}_1$ parallel to $\mathbf{r}_2$. In the first case the curve $\mathbf{R} = \mathbf{r}_1(t)$ reduces to a single point, as $\mathbf{r}_1$ is constant, and so we have a cone, which is developable. In the second case $\mathbf{r}_2$ lies along the tangent to the curve just mentioned, and so $\mathbf{R}$ takes the form $\mathbf{r}_1(t) + v\dot{\mathbf{r}}_1(t)$, where v is a new parameter, and the surface becomes the locus of tangents to a curve, and so a developable.

Theorem 3.18. *A necessary and sufficient condition for a ruled surface* $\mathbf{R} = \mathbf{r}_1(t) + u\mathbf{r}_2(t)$ *to be developable is that* $[\dot{\mathbf{r}}_1, \mathbf{r}_2, \dot{\mathbf{r}}_2] = 0$.

The condition is necessary, because for a developable surface we must have $\mathbf{r}_2$ parallel to $\dot{\mathbf{r}}_1$, or $\mathbf{r}_2 =$ constant, giving $\dot{\mathbf{r}}_2 = \mathbf{0}$. In either case the triple scalar product is zero.

The condition is sufficient, for it implies that (i) $\dot{\mathbf{r}}_1 = \mathbf{0}$, or (ii) $\dot{\mathbf{r}}_2 = \mathbf{0}$, giving respectively a cone or a cylinder, or that (iii) $\dot{\mathbf{r}}_1$, $\mathbf{r}_2$, $\dot{\mathbf{r}}_2$ are parallel to the same plane, and so we may put $\mathbf{r}_2 = \lambda(t,u)\dot{\mathbf{r}}_1 + \mu(t,u)\dot{\mathbf{r}}_2$, where $\lambda(t,u)$, $\mu(t,u)$ are scalar functions of t and u. The normal at $P(t,u)$ is then parallel to

$$\begin{aligned}(\dot{\mathbf{r}}_1 + u\dot{\mathbf{r}}_2) \wedge \mathbf{r}_2 &= (\dot{\mathbf{r}}_1 + u\dot{\mathbf{r}}_2) \wedge (\lambda\dot{\mathbf{r}}_1 + \mu\dot{\mathbf{r}}_2) \\ &= (\mu - \lambda u)\dot{\mathbf{r}}_1 \wedge \dot{\mathbf{r}}_2,\end{aligned}$$

a vector parallel to $\dot{\mathbf{r}}_1 \wedge \dot{\mathbf{r}}_2$, a function of t alone. Thus the tangent planes to S form a one-parameter family, and so S is a developable surface.

Special Families of Curves on a Surface

We have already mentioned that on any surface there are certain families of curves with special properties. We shall now prove some theorems which stress the geometrical character of these curves.

Lines of Curvature

Theorem 3.19. *A curve C on a surface S is a line of curvature if and only if the normals to S along C form a developable surface Σ.*

We have already shown in theorem 3.8 that the normals along a line of curvature form a developable surface. So it only remains to show that if the normals to S along C do form a developable surface, then C is a line of curvature.

Now C may be defined by expressing the parameters u, v of S in terms of a single new parameter t, writing $u = u(t)$, $v = v(t)$. If $\mathbf{e}$ is a unit normal to S at the point P given by $\mathbf{r}(u,v)$ on C, then a necessary and sufficient condition for the normals to S along C to form a developable surface is $[\mathbf{e},\dot{\mathbf{e}},\dot{\mathbf{r}}] = 0$, the derivatives being with respect to t. For this implies that 'consecutive' normals 'intersect', and so the ruled surface formed by them can be developed into a plane (see theorem 3.7). This condition may be written in the form

$$0 = [\mathbf{e},\dot{\mathbf{e}},\dot{\mathbf{r}}] = [\mathbf{e},\mathbf{e}_u\dot{u} + \mathbf{e}_v\dot{v},\mathbf{r}_u\dot{u} + \mathbf{r}_v\dot{v}]$$
$$= (Adu^2 + 2Bdudv + Cdv^2)/dt^2,$$

where

$$A = [\mathbf{e},\mathbf{e}_u,\mathbf{r}_u],\ 2B = [\mathbf{e},\mathbf{e}_u,\mathbf{r}_v] + [\mathbf{e},\mathbf{e}_v,\mathbf{r}_u],$$
$$C = [\mathbf{e},\mathbf{e}_v,\mathbf{r}_v].$$

We have now obtained a first-order differential equation of the second degree in the ratio du/dv, showing that there are two families of curves with the property given, one curve from each family passing through each point P on S.

To show that the differential equation $Adu^2 + 2Bdudv + Cdv^2 = 0$ is that of the lines of curvature

$$\begin{vmatrix} dv^2 & -dvdu & du^2 \\ E & F & G \\ L & M & N \end{vmatrix} = 0,$$

we must show that

$$A\sigma = EM - FL, \quad 2B\sigma = EN - GL, \quad C\sigma = FN - GM,$$

where σ is some constant of proportionality. Now with $W = (EG - F^2)^{1/2}$ we have

$$\begin{aligned} W[\mathbf{e},\mathbf{e}_u,\mathbf{r}_u] &= [\mathbf{r}_u \wedge \mathbf{r}_v,\mathbf{e}_u,\mathbf{r}_u] = \mathbf{e}_u\cdot\{\mathbf{r}_u \wedge (\mathbf{r}_u \wedge \mathbf{r}_v)\} \\ &= \mathbf{e}_u\cdot\{\mathbf{r}_u(\mathbf{r}_u\cdot\mathbf{r}_v) - \mathbf{r}_v(\mathbf{r}_u\cdot\mathbf{r}_u)\} \\ &= -LF + ME; \end{aligned}$$

similarly

$$\begin{aligned} W\{[\mathbf{e},\mathbf{e}_u,\mathbf{r}_v] + [\mathbf{e},\mathbf{e}_v,\mathbf{r}_u]\} &= \mathbf{e}_u\cdot\{\mathbf{r}_u(\mathbf{r}_v\cdot\mathbf{r}_v) - \mathbf{r}_v(\mathbf{r}_u\cdot\mathbf{r}_v)\} + \\ &\qquad + \mathbf{e}_v\cdot\{\mathbf{r}_u(\mathbf{r}_u\cdot\mathbf{r}_v) - \mathbf{r}_v(\mathbf{r}_u\cdot\mathbf{r}_u)\} \\ &= -LG + MF - MF + NE \\ &= -LG + NE, \\ W[\mathbf{e},\mathbf{e}_v,\mathbf{r}_v] &= \mathbf{e}_v\cdot\{\mathbf{r}_u(\mathbf{r}_v\cdot\mathbf{r}_v) - \mathbf{r}_v(\mathbf{r}_u\cdot\mathbf{r}_v)\} \\ &= -MG + NF, \end{aligned}$$

and so the theorem is proved with $\sigma = -W$.

A shorter method of proving the result is as follows:

Since $[\mathbf{e},\dot{\mathbf{e}},\dot{\mathbf{r}}] = 0$ the three vectors involved satisfy a linear relation

$$\lambda\mathbf{e} + \mu\dot{\mathbf{e}} + \nu\dot{\mathbf{r}} = \mathbf{0}.$$

Taking the scalar product with $\mathbf{e}$ we have at once $\lambda = 0$, for $\mathbf{e}.\dot{\mathbf{e}} = \mathbf{e}.\dot{\mathbf{r}} = 0$, as $\mathbf{e}$ is perpendicular to both $\dot{\mathbf{e}}$ and $\dot{\mathbf{r}}$.

Hence $\mu\dot{\mathbf{e}} + \nu\dot{\mathbf{r}} = \mathbf{0}$, which may be written in the form $d\mathbf{e} + \kappa d\mathbf{r} = \mathbf{0}$, provided $\mu \neq 0$, and this is true, for otherwise we should have $\dot{\mathbf{r}} = \mathbf{0}$. C is then a line of curvature by the converse of Olinde Rodrigues' theorem, already proved as theorem 3.6.

Theorem 3.20. *The two lines of curvature through any point are in perpendicular directions, except when the point is an umbilic.*

Choose parameters on S so that the lines of curvature are $u = \text{const.}$, $v = \text{const.}$ Then their differential equation is $dudv = 0$. Hence $EM - FL = 0$, $FN - GM = 0$, from which we get

$$EMN = FLN = FNL = GML.$$

So $M(EN - GL) = 0$, giving either $EN - GL = 0$, or $M = 0$. The first of these, with the earlier two equations, gives $E/L = F/M = G/N$, the case in which we have an umbilic (definition 3.12) and all normal sections have the same curvature, and $M = 0$ gives $FL = FN = 0$, so either $F = 0$, or both L and N are zero. If $F = 0$, that is $\mathbf{r}_u.\mathbf{r}_v = 0$, the lines of curvature are perpendicular. If L, M, N are all zero, so is the curvature for all normal sections, a degenerate case of an umbilic, and in this case the surface would be a plane.

Theorem 3.21. *A necessary and sufficient set of conditions for the parametric lines to be lines of curvature is that* $F = M = 0$.

If we take the parametric lines as lines of curvature their differential equation must be $dudv = 0$. The determinant form of the differential equation for the lines of curvature shows that this will occur when $FN - GM = EM - FL = 0$ and $EN - GL \neq 0$. And, assuming L, M, N not zero, this would lead to

$$\frac{E}{L} = \frac{F}{M} = \frac{G}{N},$$

a contradiction, unless F/M is indeterminate, in which case $F = 0$, and $M = 0$. If both F and M are zero, it is at once evident that the differential equation reduces to $dudv = 0$, unless $EN - GL = 0$. The exceptions noted occur in the cases of umbilics and planes, in which case the theorem becomes meaningless.

Conjugate Directions

As we have already seen the direction conjugate to that of the tangent at any point to a curve Γ on a surface S is the direction of the limit of the line of intersection of two tangent planes to S at neighbouring points of Γ. Now if the position vector to the point P on Γ is $\mathbf{r}\{u(t),v(t)\}$, then the position vector $\mathbf{R}$ of a point on the tangent plane to S at P satisfies the equation $(\mathbf{R} - \mathbf{r}).\mathbf{e} = 0$, where $\mathbf{e}$ is a unit normal to S at P. Differentiating this equation with respect to t gives the equation of the plane through the limit of the line of intersection of the plane $(\mathbf{R} - \mathbf{r}).\mathbf{e} = 0$ with its neighbour. Thus the pair of equations

$$(\mathbf{R} - \mathbf{r}).\mathbf{e} = 0, \quad (\mathbf{R} - \mathbf{r}).\dot{\mathbf{e}} - \dot{\mathbf{r}}.\mathbf{e} = 0$$

gives a line in direction conjugate to that of $\dot{\mathbf{r}}$, the tangent direction of Γ at P. But the second equation reduces to $(\mathbf{R} - \mathbf{r}).\dot{\mathbf{e}} = 0$, since $\dot{\mathbf{r}}.\mathbf{e} = 0$, and so for the conjugate direction we get $(\mathbf{R} - \mathbf{r})$ perpendicular to both $\mathbf{e}$ and $\dot{\mathbf{e}}$, and hence

Theorem 3.22. *The direction conjugate to that of* $\dot{\mathbf{r}}$ *is that of* $\mathbf{e} \wedge \dot{\mathbf{e}}$.

Notice also that the developable surface of the tangent planes to S at points of Γ may now be written as the ruled surface

$$\mathbf{R} = \mathbf{r} + w(\mathbf{e} \wedge \dot{\mathbf{e}}),$$

where w is a second parameter.

Now the vector $\mathbf{e} \wedge \dot{\mathbf{e}}$ has direction

$$\begin{aligned}
&(\mathbf{r}_u \wedge \mathbf{r}_v) \wedge (\mathbf{e}_u\dot{u} + \mathbf{e}_v\dot{v}) \\
&\quad = [\mathbf{r}_v(\mathbf{r}_u.\mathbf{e}_u) - \mathbf{r}_u(\mathbf{r}_v.\mathbf{e}_u)]\dot{u} + [\mathbf{r}_v(\mathbf{r}_u.\mathbf{e}_v) - \mathbf{r}_u(\mathbf{r}_v.\mathbf{e}_v)]\dot{v} \\
&\quad = [-L\mathbf{r}_v + M\mathbf{r}_u]\dot{u} + [-M\mathbf{r}_v + N\mathbf{r}_u]\dot{v} \\
&\quad = \mathbf{r}_u(M\dot{u} + N\dot{v}) - \mathbf{r}_v(L\dot{u} + M\dot{v}) \\
&\quad = \mathbf{r}_u\dot{u}_1 + \mathbf{r}_v\dot{v}_1, \text{ say.}
\end{aligned}$$

So for conjugate directions du/dv, du_1/dv_1 we have

$$du_1/dv_1 = -(Mdu + Ndv)/(Ldu + Mdv),$$

that is

$$Ldudu_1 + M(dudv_1 + du_1dv) + Ndvdv_1 = 0.$$

Asymptotic Lines

Theorem 3.23. *An asymptotic line on a surface S is a curve with the property that the osculating planes of Γ are tangent planes to the surface.*

For the osculating planes to be tangent planes to S the binormals to Γ must lie along the normals to S. So the vectors $\mathbf{e}$ and $\mathbf{r}' \wedge \mathbf{r}''$ are parallel. So

$$0 = \mathbf{e} \wedge (\mathbf{r}' \wedge \mathbf{r}'') = \mathbf{r}'(\mathbf{e}.\mathbf{r}'') - \mathbf{r}''(\mathbf{e}.\mathbf{r}') = \mathbf{r}'(\mathbf{e}.\mathbf{r}''),$$

since $\mathbf{e}.\mathbf{r}' = 0$. Hence $\mathbf{e}.\mathbf{r}'' = 0$, giving

$$0 = (\mathbf{r}_u \wedge \mathbf{r}_v).(\mathbf{r}_{uu}u'^2 + 2\mathbf{r}_{uv}u'v' + \mathbf{r}_{vv}v'^2 + \mathbf{r}_uu'' + \mathbf{r}_vv''),$$

and as $[\mathbf{r}_u,\mathbf{r}_v,\mathbf{r}_u] = [\mathbf{r}_u,\mathbf{r}_v,\mathbf{r}_v] = 0$, we get

$$0 = Ldu^2 + 2Mdudv + Ndv^2,$$

the equation of the asymptotic lines, which, from theorem 3.22 are seen at once to have self-conjugate directions.

It is easily verified that with any parameter t the vector $\dot{\mathbf{r}} \wedge \ddot{\mathbf{r}}$ is parallel to the vector $\mathbf{r}' \wedge \mathbf{r}''$, so we have

Theorem 3.24. *A curve Γ lying on a surface is an asymptotic line if*

$$\mathbf{e} \wedge (\dot{\mathbf{r}} \wedge \ddot{\mathbf{r}}) = \mathbf{0}.$$

Using the self-conjugate property of asymptotic lines we have $\dot{\mathbf{r}}$ parallel to $\mathbf{e} \wedge \dot{\mathbf{e}}$ and so

Theorem 3.25. *A curve Γ on a surface is an asymptotic line if*

$$\dot{\mathbf{r}} \wedge (\mathbf{e} \wedge \dot{\mathbf{e}}) = \mathbf{0},$$

and expanding this expression we get $\mathbf{0} = \mathbf{e}(\dot{\mathbf{r}}.\dot{\mathbf{e}}) - \dot{\mathbf{e}}(\dot{\mathbf{r}}.\mathbf{e})$, that is $\dot{\mathbf{r}}.\dot{\mathbf{e}} = 0$, since $\dot{\mathbf{r}}.\mathbf{e} = 0$, which gives at once

$$-(\mathbf{r}_u\dot{u} + \mathbf{r}_v\dot{v}).(\mathbf{e}_u\dot{u} + \mathbf{e}_v\dot{v}) = L\dot{u}^2 + 2M\dot{u}\dot{v} + N\dot{v}^2 = 0,$$

that is

$$Ldu^2 + 2Mdudv + Ndv^2 = 0,$$

as before.

From the geometrical point of view we see that the tangent planes to the surface along an asymptotic line Γ form a developable surface whose generators are themselves tangents to Γ, and so this surface is the developable surface of the osculating planes of Γ, these osculating planes being themselves tangent planes to the surface on which Γ lies.

Lines of Curvature

Using the idea of conjugate directions we may give

Theorem 3.26. *Orthogonal conjugate directions through a point on a surface give the directions of the lines of curvature* *through the point, and a curve Γ on the surface is a line of curvature if at every one of its points*

$$[\dot{\mathbf{r}},\mathbf{e},\dot{\mathbf{e}}] = 0.$$

For conjugate directions to be perpendicular the vector $\dot{\mathbf{r}}$ must be perpendicular to the vector $\mathbf{e} \wedge \dot{\mathbf{e}}$, so we have, as in theorem 3.22

$$0 = \dot{\mathbf{r}}.(\mathbf{e} \wedge \dot{\mathbf{e}}) = (\mathbf{r}_u\dot{u} + \mathbf{r}_v\dot{v}) . (\{- \mathbf{r}_vL + \mathbf{r}_uM\}\dot{u} + \\ + \{- \mathbf{r}_vM + \mathbf{r}_uN\}\dot{v})W,$$

that is

$$\dot{u}^2(- \mathbf{r}_u.\mathbf{r}_vL + \mathbf{r}_u.\mathbf{r}_uM) + \dot{u}\dot{v}(- \mathbf{r}_v.\mathbf{r}_vL + \mathbf{r}_u.\mathbf{r}_uN) + \\ + \dot{v}^2(- \mathbf{r}_v.\mathbf{r}_vM + \mathbf{r}_v.\mathbf{r}_uN) = 0,$$

which gives the differential equation for the lines of curvature

$$du^2(- FL + EM) + dudv(- GL + EN) + \\ + dv^2(- GM + FN) = 0.$$

Geodesics

Curves which give paths of least length joining pairs of points on a surface are of special interest. They correspond to the straight lines of a plane; they are also of interest in applied mathematics in connection with the motion of particles.

A practical method of obtaining such a curve would be to pass a string over the surface through the two points, and then draw it taut. If we regard the surface as smooth, it is then clear from the principles of statics that the principal normals to the curve formed by the string must also be normal to the surface.

We give the following

Definition 3.19. *If two points A, B are taken on a surface S and joined by curves Γ lying on S, then any such curve Γ_0 possessing a stationary length for small variations over the surface is called a* **geodesic.**

For example, if A, B lie on the surface of a sphere, but are not the extremities of a diameter, then the great circle passing through A, B is a geodesic. It will be divided into two portions, one having the shortest, and the other the longest distance between A and B; but both portions have stationary values for their lengths when the curve is considered as one of a whole set of curves which join A and B.

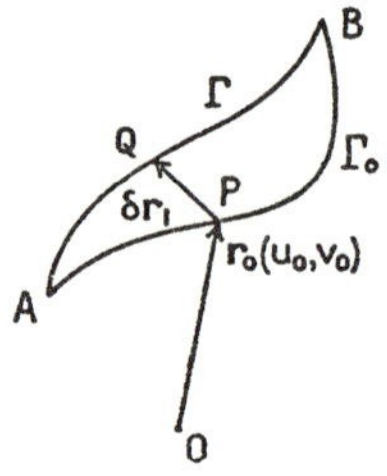

FIG. 3.9

We shall now find the condition for a curve joining two points on the surface to have a stationary value for its length.

Suppose that Γ_0 is a geodesic passing through two points A and B on a surface S, then we write the position vector to any point P on Γ_0 as $\mathbf{r}_0(u_0,v_0)$, where u_0, v_0 are each specified functions of a parameter t determining the point P. Let us vary the curve Γ_0 to Γ, a curve still passing through A and B, by giving each point P a small displacement $\delta\mathbf{r}_1$ to a point Q on S, where $\delta\mathbf{r}_1$ is an arbitrary function of t taking zero values at both A and B. If g, g_0 are the lengths from A to B of Γ, Γ_0 then g_0 will have a stationary value if the difference $\delta g_0 = g - g_0$ is a magnitude of the second order of smallness. Now

$$\delta g_0 = \int_A^B \{[\dot{\overline{(\mathbf{r}_0 + \delta\mathbf{r}_1)}} \cdot \dot{\overline{(\mathbf{r}_0 + \delta\mathbf{r}_1)}}]^{1/2} - [\dot{\mathbf{r}}_0 \cdot \dot{\mathbf{r}}_0]^{1/2}\}\, dt$$

$$= \int_A^B (\dot{\mathbf{r}}_0.\dot{\mathbf{r}}_0)^{1/2} \left\{ \left[1 + \frac{2\dot{\mathbf{r}}_0.\dot{\overline{\delta \mathbf{r}_1}} + \dot{\overline{\delta \mathbf{r}_1}}.\dot{\overline{\delta \mathbf{r}_1}}}{(\dot{\mathbf{r}}_0.\dot{\mathbf{r}}_0)} \right]^{1/2} - 1 \right\} dt$$

$$= \int_A^B \frac{(\dot{\mathbf{r}}_0.\dot{\overline{\delta \mathbf{r}_1}})}{(\dot{\mathbf{r}}_0.\dot{\mathbf{r}}_0)^{1/2}}\, dt + \text{terms of order at least that of } (\dot{\overline{\delta \mathbf{r}_1}})^2.$$

Integrating by parts we get

$$\delta g_0 = \left[\delta \mathbf{r}_1 \cdot \frac{\dot{\mathbf{r}}_0}{(\dot{\mathbf{r}}_0.\dot{\mathbf{r}}_0)^{1/2}} \right]_A^B - \int_A^B \delta \mathbf{r}_1 \cdot \frac{d}{dt}\left[\frac{\dot{\mathbf{r}}_0}{(\dot{\mathbf{r}}_0.\dot{\mathbf{r}}_0)^{1/2}} \right] dt + \ldots .$$

The first term on the right vanishes since $\delta \mathbf{r}_1$ is zero at both A and B, and so for δg_0 to be at least of the second order of smallness we must have

$$\int_A^B \delta \mathbf{r}_1 \cdot \frac{d}{dt}\left[\frac{\dot{\mathbf{r}}_0}{(\dot{\mathbf{r}}_0.\dot{\mathbf{r}}_0)^{1/2}} \right] dt = 0$$

for all variations $\delta \mathbf{r}_1$. The only way in which this can happen is for the integrand to be zero for all values of t.

So the condition for a geodesic is that for all t

$$\delta \mathbf{r}_1 \cdot \frac{d}{dt}\left[\frac{\dot{\mathbf{r}}_0}{(\dot{\mathbf{r}}_0.\dot{\mathbf{r}}_0)^{1/2}} \right] = 0.$$

To obtain the geometrical significance of this result let us choose the arc-length s ($\equiv AP_0$) of Γ_0 as our parameter t. Then $|\dot{\mathbf{r}}_0| = |\mathbf{r}_0'| = 1$, and so $\dot{\mathbf{r}}_0.\dot{\mathbf{r}}_0 = 1$ and our result becomes

$$\delta \mathbf{r}_1.\mathbf{r}''_0 = 0$$

for all values of s.

As $\delta\mathbf{r}_1$ is an arbitrary vector, the displacement from P to Q on S, if we divide by δs_1, the arc-length from P to Q, and take the limit as $\delta s_1 \to 0$, leads to an arbitrary tangent vector, $d\mathbf{r}_1/ds_1$, to S at P, and the result $\mathbf{r}_0''.(d\mathbf{r}_1/ds_1) = 0$ shows that $\mathbf{r}_0''$ is perpendicular to any tangent vector to S, unless $\mathbf{r}_0'' = \mathbf{0}$, in which case $\mathbf{r}_0 = \mathbf{a}s + \mathbf{b}$, and Γ_0 is a straight line. So we get

Theorem 3.27. *The principal normal at any point P on Γ, a curved geodesic on S, is also normal to S at P.*

It follows at once that the binormal to Γ is perpendicular to $\mathbf{e}$, and so $\mathbf{b}.\mathbf{e} = 0$. But $\mathbf{b} = \mathbf{t} \wedge \mathbf{p} = \mathbf{r}' \wedge \mathbf{r}''/\kappa$, giving $[\mathbf{r}',\mathbf{r}'',\mathbf{e}] = 0$.

Remembering that $\mathbf{r}' \wedge \mathbf{r}''$ is a vector parallel to $\dot{\mathbf{r}} \wedge \ddot{\mathbf{r}}$ we may give

Theorem 3.28. *For a geodesic on a surface*

$$[\mathbf{e},\dot{\mathbf{r}},\ddot{\mathbf{r}}] = 0.$$

If we expand this condition we get

$$\begin{aligned} 0 &= \dot{\mathbf{r}}.[\ddot{\mathbf{r}} \wedge (\mathbf{r}_u \wedge \mathbf{r}_v)] \\ &= (\dot{\mathbf{r}}.\mathbf{r}_u)(\ddot{\mathbf{r}}.\mathbf{r}_v) - (\dot{\mathbf{r}}.\mathbf{r}_v)(\ddot{\mathbf{r}}.\mathbf{r}_u) \\ &= [(\mathbf{r}_u\dot{u} + \mathbf{r}_v\dot{v}).\mathbf{r}_u][\mathbf{r}_{uu}\dot{u}^2 + 2\mathbf{r}_{uv}\dot{u}\dot{v} + \\ &\qquad\qquad + \mathbf{r}_{vv}\dot{v}^2 + \mathbf{r}_u\ddot{u} + \mathbf{r}_v\ddot{v}).\mathbf{r}_v] \\ &\quad - [(\mathbf{r}_u\dot{u} + \mathbf{r}_v\dot{v}).\mathbf{r}_v][(\mathbf{r}_{uu}\dot{u}^2 + 2\mathbf{r}_{uv}\dot{u}\dot{v} + \\ &\qquad\qquad + \mathbf{r}_{vv}\dot{v}^2 + \mathbf{r}_u\ddot{u} + \mathbf{r}_v\ddot{v}).\mathbf{r}_u]. \end{aligned}$$

This may be put in the form of

Theorem 3.29. *A differential equation giving the geodesics on a surface is*

$$\begin{aligned} &[E\dot{u} + F\dot{v}][\mathbf{r}_v.(\mathbf{r}_{uu}\dot{u}^2 + 2\mathbf{r}_{uv}\dot{u}\dot{v} + \mathbf{r}_{vv}\dot{v}^2) + F\ddot{u} + G\ddot{v})] \\ &- [F\dot{u} + G\dot{v}][\mathbf{r}_u.(\mathbf{r}_{uu}\dot{u}^2 + 2\mathbf{r}_{uv}\dot{u}\dot{v} + \mathbf{r}_{vv}\dot{v}^2) + E\ddot{u} + F\ddot{v})] = 0. \end{aligned}$$

By writing $\dot{u}/\dot{v} = du/dv$ we see that this is a second order differential equation, and the values of the constants will be determined by the values of (u,v) and du/dv giving

Theorem 3.30. *Through any point on a surface there is in general just one geodesic in any given direction.*

Geodesic Curvature

For any curve Γ to the first order of small quantities

$$\delta g = -\int_A^B \delta\mathbf{r}_1.\mathbf{r}''\, ds,$$

where $\delta\mathbf{r}_1$ represents the displacement from a point P on the curve to another point Q on the surface.

The integrand is $\delta\mathbf{r}_1.\kappa\mathbf{p}$, whose modulus is $|\delta\mathbf{r}_1|\kappa \cos\phi$, where ϕ is the angle between $\mathbf{p}$ and $\delta\mathbf{r}_1$. If θ is the angle between $\mathbf{p}$ and $\mathbf{e}$, and we take our $\delta\mathbf{r}_1$ so that the limiting position of a vector along $\delta\mathbf{r}_1$ is perpendicular to Γ, then $\cos\phi$ will approach $\pm \sin\theta$. We then give

Definition 3.20. *κ sin θ is called the* **geodesic curvature** *of Γ.*

The name is due to the fact that it becomes zero when Γ is actually a geodesic.

Theorem 3.31. *Geodesic curvature is an intrinsic property of a curve on a surface.*

This follows at once since δg, $\delta\mathbf{r}_1$ and the lengths of pairs of corresponding curves on two isometric surfaces remain the same. So $\kappa \sin\theta$ is constant for any isometric deformation of Γ.

EXAMPLES ON CHAPTER THREE

3.1. A curve Γ on a surface S cuts all the parametric curves $u =$ const., at a constant angle α. If the two sets of parametric curves $u =$ const., $v =$ const. are orthogonal, show that the differential equation of Γ is $E^{1/2} \cos\alpha.du = G^{1/2} \sin\alpha.dv$.

3.2. Find in terms of u, v the position vector **r** of the point (x,y,z) on the **hyperbolic paraboloid** whose equations are

$$bx + ay = 2abu, \quad bx - ay = 2abv, \quad z = 2cuv.$$

What are the parametric curves on this surface, and what is the x,y,z equation of the surface?

3.3. Find E, F, G for the **sphere** given by

$$x = a\cos\theta\sin\phi, \quad y = a\sin\theta\sin\phi, \quad z = a\cos\phi.$$

Find the lengths of the curves on the surface of the sphere given by

(i) $\theta = \alpha$, from $\phi = 0$ to $\phi = \pi$;

(ii) $\phi = \beta$, from $\theta = 0$ to $\theta = 2\pi$;

(iii) $\theta = \log(\sec\phi + \tan\phi)$ from $\phi = 0$ to $\phi = \pi/4$.

Calculate the surface area A of the sphere from the expression

$$dA = (EG - F^2)^{1/2}\, d\theta d\phi.$$

3.4. Give a geometrical description of the **right helicoid** given by

$$x = u\cos\theta, \quad y = u\sin\theta, \quad z = c\theta,$$

and show that the curves given by $du^2 = (u^2 + c^2)d\theta^2$ form an orthogonal net on the surface.

3.5. Show that the curves $u = a$, $u = b$ on the surface given by

$$\mathbf{r} = \mathbf{i}u\cos\theta + \mathbf{j}u\sin\theta + \mathbf{k}(\theta + \log\cos u)$$

make intercepts of the same length on the curves $\theta = \lambda$ for all values of λ.

3.6. Prove **Euler's theorem** that the curvature κ_N of a curve Γ, lying in a normal section of a surface S, at a point P, where the tangent to Γ makes an angle α with one of the lines of curvature of S through P, is given in terms of the principal curvatures K_1, K_2 at P by the relation $\kappa_N = K_1\cos^2\alpha + K_2\sin^2\alpha$.

3.7. Prove **Dupin's theorem** that the sum of the normal curvatures in two perpendicular directions is constant.

3.8. For the **surface of revolution** given by

$$x = u\cos\theta, \quad y = u\sin\theta, \quad z = f(u),$$

give a geometrical interpretation of u, and find an integral giving the length of a curve on the surface.

Determine also the geometrical nature of the lines of curvature.

3.9. Give a geometrical description of the **catenoid** given by

$$u = c\cosh(z/c), \quad u^2 = x^2 + y^2,$$

and find an expression in terms of u, θ, where $x = u\cos\theta$, $y = u\sin\theta$, for **r** the position vector of a point on the surface. Determine the lines of curvature and the asymptotic lines, and show that the latter form an orthogonal net.

3.10. The **Third Fundamental Quadratic Form** for a surface is defined to be $d\mathbf{e}.d\mathbf{e}$. Denoting the three quadratic forms by I, II, III, show that

$$(K_1K_2)\mathrm{I} - (K_1 + K_2)\mathrm{II} + \mathrm{III} = 0.$$

3.11. Show that the torsions of the asymptotic lines through a point P are $\pm(-K)^{1/2}$, where K is the Gauss curvature of the surface at P.

3.12. Find an equation giving the principal curvatures K_1, K_2 for the ruled surface formed by the normals to a curve Γ given by

$$\mathbf{R} = \mathbf{r}(s) + u\mathbf{p}(s),$$

where u is a parameter.

3.13. Show that the asymptotic lines through a point on a surface have directions bisected by the directions of the lines of curvature through the point.

3.14. If a surface S cuts a plane π everywhere at the same angle, show that the section of S by π is a line of curvature on S.

3.15. On a particular surface S the element of arc ds is given by $ds^2 = du^2 + \lambda^2 dv^2$, where λ is a function of both u and v. Prove that $K = -\partial^2\lambda/\lambda\partial u^2$, and apply this result to the sphere of question 3.3.

3.16. For the helicoid of question 3.4 show that the asymptotic lines are the parametric curves. Find the equations of the lines of curvature as relations between u and θ, and show that the principal radii of curvature are equal in magnitude, but opposite in sign, and independent of the parameter v.

3.17. If the normals to the surface $2z = K_1x^2 + K_2y^2$ at the points $(0,0,0)$ and (α,β,γ), a neighbouring point, intersect, show that (α,β,γ) lies on a line of curvature through the origin.

3.18. Find equations for the asymptotic lines and the lines of curvature for the surface of question 3.2.

3.19. For a surface S given in **Monge's** form $z = f(x,y)$ calculate $E, F, G, L, M, N, \mathbf{e}, K$ in terms of p, q, r, s, t where

$$p = \partial z/\partial x, \quad q = \partial z/\partial y, \quad r = \partial^2 z/\partial x^2, \quad s = \partial^2 z/\partial x \partial y, \quad t = \partial^2 z/\partial y^2.$$

Show that the condition for S to be developable is that $rt = s^2$, (i) from the evaluation of K, (ii) by finding the equation of the tangent plane to S and showing it to be a member of a one-parameter family.

3.20. Find the equations of the lines of curvature on Σ, the developable surface of tangents to a curve Γ given by $\mathbf{R} = \mathbf{r}(s) + u\mathbf{t}(s)$, and show that the asymptotic lines are the generators of Σ.

3.21. For a certain surface S we are given that $\mathbf{e}_u$ is parallel to $\mathbf{e}_v$. By taking new coordinate curves $u' =$ const., $v' =$ const., one set of which are the asymptotic lines of S, show that either $\mathbf{e}_{u'}$, or $\mathbf{e}_{v'}$, is zero, and hence that S is developable.

3.22. Show that the edge of regression of the developable surface enveloped by the normal planes of a curve Γ is the locus of the centres of spherical curvature of Γ.

3.23. Show that the torsion λ of a geodesic is given by

$$\lambda = [\mathbf{e},\mathbf{e}',\mathbf{r}'].$$

3.24. Show that the geodesic curvature κ_G of a curve Γ is given by

$$\kappa_G = [\mathbf{e},\mathbf{r}',\mathbf{r}''].$$

3.25. If S is the ruled surface generated by the binormals of a curve Γ show that Γ is a geodesic on S.

SOLUTIONS TO EXAMPLES

Chapter One

1.1. $|\kappa\mathbf{p}| = |d\mathbf{t}/ds|$, $|\kappa\mathbf{t}\wedge\mathbf{p}| = |\mathbf{t}\wedge d\mathbf{t}/ds|$, $|\kappa| = |(\mathbf{i}x'+\mathbf{j}y')\wedge(\mathbf{i}x''+\mathbf{j}y'')| = |\mathbf{k}(x'y''-x''y')| = |y''x'-x''y'|$.

1.2. $|\kappa| = |\mathbf{t}\wedge d\mathbf{t}/ds| = |(\dot{\mathbf{r}}/\dot s)\wedge(\dot{\mathbf{t}}/\dot s)|$

$= |[(\mathbf{i}\dot x+\mathbf{j}\dot y)/\dot s]\wedge[(\mathbf{i}\ddot x+\mathbf{j}\ddot y)\dot s-(\mathbf{i}\dot x+\mathbf{j}\dot y)\ddot s]/\dot s^3| = |(\dot x\ddot y-\dot y\ddot x)/\dot s^3|$

$= |(\dot x\ddot y-\dot y\ddot x)/(\dot x^2+\dot y^2)^{3/2}|$.

1.3. If ϕ is the angle between radius vector and tangent, and $\mathbf{r} = r\hat{\mathbf{r}}$, where $\hat{\mathbf{r}}$ is a unit vector, then $r\cos\phi = \mathbf{r}.\mathbf{t} = \mathbf{r}.(d\mathbf{r}/ds) = \mathbf{r}.\hat{\mathbf{r}}(dr/ds) + \mathbf{r}.r\,(d\hat{\mathbf{r}}/ds) = (\hat{\mathbf{r}}.\hat{\mathbf{r}})r(dr/ds) = r(dr/ds)$. So $\cos\phi = dr/ds$.

1.4. If ϕ is the angle between radius vector and tangent $p = r\sin\phi = |\mathbf{r}.\mathbf{n}|$. $|dp/ds| = |\mathbf{n}.d\mathbf{r}/ds + \mathbf{r}.d\mathbf{n}/ds| = |\mathbf{r}.\mathbf{t}\,d\psi/ds|$. $|\kappa| = |(dp/ds)/(\mathbf{r}.\mathbf{t})| = |(dp/ds)/r\cos\phi| = |dp/r\,dr|$.

1.5. By Pythagoras from geometrical significance of $(dp/d\psi)$, or at once from $\mathbf{r} = \mathbf{n}p + \mathbf{m}\,(dp/d\psi)$.

1.6. If ψ is the angle made by the tangent with the perpendicular from O to the fixed line $a = p\cos\phi = p\sin\psi$, $dp/d\psi = -a\cos\psi/\sin^2\psi$, $p^2 + (dp/d\psi)^2 = (a^2/\sin^2\psi) + (a^2\cos^2\psi/\sin^4\psi) = (a^2/\sin^4\psi) = r^2$, $2a = 2r\cos^2\phi = r(1+\cos 2\phi)$, $x = a - r\cos 2\phi = a-(2a-r)$, $(x+a)^2 = r^2 = y^2+(a-x)^2$, $y^2 = 4ax$.

Chapter Two

2.1. $\mathbf{t} = (3\mathbf{i}a\theta^2 + 2\mathbf{j}b\theta + \mathbf{k}c)/(9a^2\theta^4 + 4b^2\theta^2 + c^2)^{1/2}.(x - a\theta^3)/3a\theta^2 = (y-b\theta^2)/2b\theta = (z-c\theta)/c$.

2.2. $\ddot{\mathbf{r}} = 0$, $\dot{\mathbf{r}} = \mathbf{c}$, $\mathbf{r} = \mathbf{c}t + \mathbf{b}$, a straight line, or the point $\mathbf{b}$ if $t = 0$.

2.3. $\dddot{\mathbf{r}} = 0$, $\ddot{\mathbf{r}} = \mathbf{c}$, $\dot{\mathbf{r}} = \mathbf{c}t + \mathbf{b}$, $\mathbf{r} = \frac{1}{2}\mathbf{c}t^2 + \mathbf{b}t + \mathbf{a}$. $\mathbf{c} \neq 0$, $\mathbf{b} \neq 0$, a parabola; $\mathbf{c} \neq 0$, $\mathbf{b} = 0$, a semi-infinite segment of a line; $\mathbf{c} = 0$, $\mathbf{b} \neq 0$, a straight line; $\mathbf{b} = 0$, $\mathbf{c} = 0$, a point.

2.4. $\mathbf{t} = \dot{\mathbf{r}}/\dot s$, $\mathbf{p} = (\ddot{\mathbf{r}}\dot s - \dot{\mathbf{r}}\ddot s)/\kappa\dot s^3$, $\mathbf{b} = (\dot{\mathbf{r}}\wedge\ddot{\mathbf{r}})/\kappa\dot s^3$, $\kappa = |(\dot{\mathbf{r}}\wedge\ddot{\mathbf{r}})|/\dot s^3$.

2.5. $\mathbf{b}$ is parallel to $\dot{\mathbf{r}}\wedge\ddot{\mathbf{r}}$. Spin-vector of $\mathbf{b} = \lambda\mathbf{t} = (\dot{\mathbf{r}}\wedge\ddot{\mathbf{r}})\wedge(\dot{\mathbf{r}}\wedge\dddot{\mathbf{r}})/(\dot{\mathbf{r}}\wedge\ddot{\mathbf{r}})^2 = \dot{\mathbf{r}}[\dot{\mathbf{r}},\ddot{\mathbf{r}},\dddot{\mathbf{r}}]/\kappa^2\dot s^6$, so $\kappa^2\lambda = [\dot{\mathbf{r}},\ddot{\mathbf{r}},\dddot{\mathbf{r}}]/\dot s^6$.

2.6. (i) $\mathbf{t} = \{3(1 - t^2)a, 6at, 3a(1 + t^2)\}/3a(2 + 4t^2 + 2t^4)^{1/2} = (1 - t^2, 2t, 1 + t^2)/\sqrt{2}(1 + t^2)$. $\dot{\mathbf{r}} = \{3(1 - t^2)a, 6at, 3a(1 + t^2)\}$, $\dot{s} = \sqrt{2}(1 + t^2)3a$. $\ddot{\mathbf{r}} = (-6at, 6a, 6at)$, $\dddot{\mathbf{r}} = (-6a, 0, 6a)$. $\kappa = |\dot{\mathbf{r}} \wedge \ddot{\mathbf{r}}|/\dot{s}^3 = [(18a^2t^2 - 18a^2)^2 + (-36a^2t)^2 + (18a^2t^2 + 18a^2)^2]^{1/2}/2\sqrt{2}(1 + t^2)^3 27a^3 = 1/3a(1 + t^2)^2$. $\lambda = [-6a(18a^2t^2 - 18a^2) + 6a(18a^2t^2 + 18a^2)]9a^2(1 + t^2)^4/8(1 + t^2)^6 3^6 a^6 = 1/3a(1 + t^2)^2$.

(ii) $C = \cosh t$, $S = \sinh t$, $C^* = \cosh 2t$. $\mathbf{t} = (aS, aC, c)/(a^2C^* + c^2)^{1/2}$, $\dot{\mathbf{r}} = (aS, aC, c)$, $\dot{s} = (a^2C^* + c^2)^{1/2}$, $\ddot{\mathbf{r}} = (aC, aS, 0)$, $\dddot{\mathbf{r}} = (aS, aC, 0)$. $\kappa = [(-acS)^2 + (acC)^2 + (-a^2)^2]^{1/2}/(a^2C^* + c^2)^{3/2} = a(c^2C^* + a^2)^{1/2}/(a^2C^* + c^2)^{3/2}$. $\lambda = (-a^2cS^2 + a^2cC^2)(a^2C^* + c^2)^3/(a^2C^* + c^2)^3a^2(c^2C^* + a^2) = c/(c^2C^* + a^2)$.

(iii) $\mathbf{t} = \{a(1 - \cos\theta), a\sin\theta, b\}/\{a^2(2 - 2\cos\theta) + b^2\}^{1/2}$, $\dot{\mathbf{r}} = \{a(1 - \cos\theta), a\sin\theta, b\}$, $\dot{s} = (a^2 4\sin^2 \tfrac{1}{2}\theta + b^2)^{1/2}$, $\ddot{\mathbf{r}} = (a\sin\theta, a\cos\theta, 0)$, $\dddot{\mathbf{r}} = (a\cos\theta, -a\sin\theta, 0)$, $\kappa = [(-ab\cos\theta)^2 + (ab\sin\theta)^2 + (a^2\cos\theta - a^2)^2]^{1/2}/\left(a^2 4\sin^2\frac{\theta}{2} + b^2\right)^{3/2} = a\left(b^2 + 4a^2\sin^2\frac{\theta}{2}\right)^{1/2}/(4a^2\sin^2\frac{\theta}{2} + b^2)^{3/2}$ $= a/\left(4a^2\sin^2\frac{\theta}{2} + b^2\right)$, $\lambda = (-a^2b\cos^2\theta - a^2b\sin^2\theta)/a^2(4a^2\sin^2\frac{\theta}{2} + b^2)$ $= -b/\left(4a^2\sin^2\frac{\theta}{2} + b^2\right)$.

2.7. $\mathbf{u}.\mathbf{u} = 1$, $\mathbf{u}.\dot{\mathbf{u}} = 0$, etc., $\mathbf{v}.\mathbf{w} = 0$, etc., $\mathbf{u} = \mathbf{v} \wedge \mathbf{w}$. $[\dot{\mathbf{u}}, \dot{\mathbf{v}}, \dot{\mathbf{w}}] = \dot{\overline{\mathbf{v} \wedge \mathbf{w}}}.(\dot{\mathbf{v}} \wedge \dot{\mathbf{w}}) = (\dot{\overline{\mathbf{v} \wedge \mathbf{w}}} \wedge \dot{\mathbf{v}}).\dot{\mathbf{w}} = [(\dot{\mathbf{v}} \wedge \mathbf{w} + \mathbf{v} \wedge \dot{\mathbf{w}}) \wedge \dot{\mathbf{v}}].\dot{\mathbf{w}} = (\dot{\mathbf{w}}.\mathbf{w})(\dot{\mathbf{v}}.\dot{\mathbf{v}}) - (\dot{\mathbf{w}}.\dot{\mathbf{v}})(\mathbf{w}.\dot{\mathbf{v}}) + (\dot{\mathbf{w}}.\dot{\mathbf{w}})(\mathbf{v}.\dot{\mathbf{v}}) - (\dot{\mathbf{w}}.\mathbf{v})(\dot{\mathbf{w}}.\dot{\mathbf{v}}) = (-\dot{\mathbf{w}}.\dot{\mathbf{v}})(\dot{\overline{\mathbf{w}.\mathbf{v}}}) = 0$.

2.8. $\mathbf{t}'.\mathbf{p}' = \kappa\mathbf{p}.\mathbf{p}' = 0$, $\mathbf{p}'.\mathbf{b}' = (-\kappa\mathbf{t} + \lambda\mathbf{b}).(-\lambda\mathbf{p}) = 0$, $\mathbf{b}'.\mathbf{t}' = (-\lambda\mathbf{p}).(\kappa\mathbf{p}) = -\lambda\kappa$. $\mathbf{t}'.(\lambda\mathbf{t} + \kappa\mathbf{b}) = \kappa\mathbf{p}.(\lambda\mathbf{t} + \kappa\mathbf{b}) = 0$, $\mathbf{p}'.(\lambda\mathbf{t} + \kappa\mathbf{b}) = (-\kappa\mathbf{t} + \lambda\mathbf{b}).(\lambda\mathbf{t} + \kappa\mathbf{b}) = -\kappa\lambda + \kappa\lambda = 0$, $\mathbf{b}'.(\lambda\mathbf{t} + \kappa\mathbf{b}) = -\lambda\mathbf{p}.(\lambda\mathbf{t} + \kappa\mathbf{b}) = 0$.

2.9. The three principal planes are $[\mathbf{R} - \mathbf{r}, \mathbf{p}, \mathbf{b}] = 0$, $[\mathbf{R} - \mathbf{r}, \mathbf{b}, \mathbf{t}] = 0$, $[\mathbf{R} - \mathbf{r}, \mathbf{t}, \mathbf{p}] = 0$. That is $(\mathbf{R} - \mathbf{r}).\mathbf{t} = 0$, $(\mathbf{R} - \mathbf{r}).\mathbf{p} = 0$, $(\mathbf{R} - \mathbf{r}).\mathbf{b} = 0$. In cartesian coordinates these become $(X - x)\dot{x} + (Y - y)\dot{y} + (Z - z)\dot{z} = 0$, $\Sigma(X - x)(\ddot{x}\dot{s} - \dot{x}\ddot{s}) = 0$, or $\Sigma(X - x)[(\ddot{x}(\dot{z}^2 + \dot{y}^2) - \dot{x}(\dot{z}\ddot{z} + \dot{y}\ddot{y})] = 0$, and $\Sigma(X - x)(\dot{y}\ddot{z} - \ddot{y}\dot{z}) = 0$.

2.10. For the sphere $(\mathbf{r} - \mathbf{c}).(\mathbf{r} - \mathbf{c}) = R^2$ to have four point contact with the curve we must have $\mathbf{t}.(\mathbf{r} - \mathbf{c}) = 0$, $\kappa\mathbf{p}.(\mathbf{r} - \mathbf{c}) + 1 = 0$, and $(\kappa'\mathbf{p} - \kappa^2\mathbf{t} + \kappa\lambda\mathbf{b}).(\mathbf{r} - \mathbf{c}) = 0$, so $-\kappa'\kappa^{-1} + \kappa\lambda\mathbf{b}.(\mathbf{r} - \mathbf{c}) = 0$. Now $(\mathbf{r} - \mathbf{c}) = u\mathbf{t} + v\mathbf{p} + w\mathbf{b}$, where $u = [(\mathbf{r} - \mathbf{c}), \mathbf{p}, \mathbf{b}]/[\mathbf{t}, \mathbf{p}, \mathbf{b}] = (\mathbf{r} - \mathbf{c}).\mathbf{t}$, etc., so $(\mathbf{r} - \mathbf{c}) = -\kappa^{-1}\mathbf{p} + \kappa'\kappa^{-2}\lambda^{-1}\mathbf{b}$ whence

$$\mathbf{c} = \mathbf{r} + \frac{\mathbf{p}}{\kappa} - \frac{\kappa'\mathbf{b}}{\kappa^2\lambda}.$$

Also $\mathbf{t} \wedge \mathbf{t}'' = \mathbf{t} \wedge [\kappa'\mathbf{p} + \kappa(-\kappa\mathbf{t} + \lambda\mathbf{b})] = \kappa'\mathbf{b} - \kappa\lambda\mathbf{p}$, so from above $R = |\mathbf{r} - \mathbf{c}| = |(\mathbf{t} \wedge \mathbf{t}'')/\kappa^2\lambda|$, and $R^2 = [(\kappa')^2 + \kappa^2\lambda^2]/\kappa^4\lambda^2 = (\rho')^2\sigma^2 + \rho^2$. For coincidence of the centre of curvature and the centre of spherical curvature $\kappa'/\kappa^2\lambda = 0$, that is $\kappa =$ constant. For R to be constant

$$\frac{d}{ds}\left[\frac{1}{\kappa^2} + \left(\frac{\kappa'}{\kappa^2\lambda}\right)^2\right] = 0,$$

that is

$$-\frac{2\kappa'}{\kappa^3} + \frac{2\kappa'}{\kappa^2\lambda}\frac{d}{ds}\left(\frac{\kappa'}{\kappa^2\lambda}\right) = 0;$$

so

$$\frac{\lambda}{\kappa} = \frac{d}{ds}\left(\frac{\kappa'}{\kappa^2\lambda}\right)$$

is the condition for a curve to lie on the surface of a sphere.

2.11. As in the case of tangents

$$d = \left|\frac{\mathbf{p} \wedge (\mathbf{p} + \delta\mathbf{p})}{|\mathbf{p} \wedge (\mathbf{p} + \delta\mathbf{p})|} \cdot \delta\mathbf{r}\right|$$

$$= \frac{|\{\mathbf{p} \wedge [(-\kappa\mathbf{t} + \lambda\mathbf{b})\delta s + \ldots]\}.\{\mathbf{t}\delta s + \ldots\}|}{|\mathbf{p} \wedge [(-\kappa\mathbf{t} + \lambda\mathbf{b})\delta s + \ldots]|}$$

$$= \frac{|\lambda\delta s^2 + \ldots|}{|(\kappa\mathbf{b} + \lambda\mathbf{t})\delta s + \ldots|} \doteqdot \frac{|\lambda\delta s|}{(\kappa^2 + \lambda^2)^{1/2}}.$$

$$\theta \doteqdot \sin\theta = \frac{|\mathbf{p} \wedge (\mathbf{p} + \delta\mathbf{p})|}{|\mathbf{p}||\mathbf{p} + \delta\mathbf{p}|} \doteqdot \frac{|\kappa\mathbf{b} + \lambda\mathbf{t}|\,\delta s}{|\mathbf{p}||\mathbf{p} + \delta\mathbf{p}|} = (\kappa^2 + \lambda^2)^{1/2}\,\delta s.$$

2.12. If the curve is a helix $\mathbf{t}.\mathbf{f} = \cos\alpha$, so $\kappa\mathbf{p}.\mathbf{f} = 0$, so $\mathbf{p}$ is perpendicular to $\mathbf{f}$ and the principal normals are all parallel. Conversely, parallel principal normals gives $\mathbf{p}.\mathbf{f} = 0$, leading to $\mathbf{t}.\mathbf{f} = \cos\alpha$.

2.13. $\mathbf{r} = \mathbf{i}a\cos\theta + \mathbf{j}a\sin\theta + \mathbf{k}a\theta\tan\alpha$, $\mathbf{t} = (-\mathbf{i}\sin\theta + \mathbf{j}\cos\theta + \mathbf{k}\tan\alpha)/\sec\alpha$, $\kappa\mathbf{p} = (-\mathbf{i}\cos\theta - \mathbf{j}\sin\theta)/a\sec^2\alpha$, so $\kappa = a^{-1}\cos^2\alpha$. $\mathbf{b} = \mathbf{t} \wedge \mathbf{p} = \kappa^{-1}(\mathbf{i}\sin\theta\tan\alpha - \mathbf{j}\cos\theta\tan\alpha + \mathbf{k})/a\sec^3\alpha$, $-\lambda\mathbf{p} = (\mathbf{i}\cos\theta\tan\alpha + \mathbf{j}\sin\theta\tan\alpha)/a\sec^2\alpha$, $\lambda = a^{-1}\tan\alpha/\sec^2\alpha = a^{-1}\sin\alpha\cos\alpha$, so both curvature and torsion are constant. $[\mathbf{t}',\mathbf{t}'',\mathbf{t}'''] = [\kappa\mathbf{p}, \kappa'\mathbf{p} + \kappa(-\kappa\mathbf{t} + \lambda\mathbf{b}), \kappa''\mathbf{p} + 2\kappa'(-\kappa\mathbf{t} + \lambda\mathbf{b}) + \kappa(-\kappa'\mathbf{t} - \kappa^2\mathbf{p} + \lambda'\mathbf{b} - \lambda^2\mathbf{p})] = \kappa(\kappa\lambda)(-3\kappa'\kappa)[\mathbf{p},\mathbf{b},\mathbf{t}] + \kappa(-\kappa^2)(2\kappa'\lambda + \kappa\lambda')[\mathbf{p},\mathbf{t},\mathbf{b}] = \kappa^5(-\kappa'\lambda + \kappa\lambda')/\kappa^2 = \kappa^5(\lambda/\kappa)'$. For a helix $\lambda/\kappa = \text{const.}$, so $[\mathbf{t}',\mathbf{t}'',\mathbf{t}'''] = [\mathbf{r}'',\mathbf{r}''',\mathbf{r}''''] = 0$. If $[\mathbf{r}'',\mathbf{r}''',\mathbf{r}''''] = 0$, $\kappa^5(\lambda/\kappa)' = 0$, so $\kappa = 0$, giving a straight line, or $\lambda/\kappa = \text{const.}$, giving a helix.

2.14. Using the results in the first part of example 2.13 above, if $\mathbf{c}$ is the vector to the centre of curvature then $\mathbf{c} = \mathbf{r} + a\sec^2\alpha\,\mathbf{p} = \mathbf{i}a\cos\theta + \mathbf{j}a\sin\theta + \mathbf{k}a\theta\tan\alpha - \mathbf{i}a\sec^2\alpha\cos\theta - \mathbf{j}a\sec^2\alpha\sin\theta = -\mathbf{i}a\tan^2\alpha\cos\theta - \mathbf{j}a\tan^2\alpha\sin\theta + \mathbf{k}\,a\theta\tan\alpha$, which gives a helix lying on a coaxial circular cylinder of radius $a\tan^2\alpha$, and so on the same cylinder if $\alpha = \pm\,\pi/4$.

2.15. If β be the semi-vertical angle of a cone of vertex V, VG and VH two generators, with G and H on the base such that GOH is a right-angle, O being the centre of the base, we take PK the tangent to the curve at P, the point where the curve cuts VG at an angle α, and K its intersection with the plane VOH. Then if Q, N are respectively the projections of P, K on the base, QN is a tangent to the projection of the curve on the cone. We get $\tan\alpha = VK/VP$, $\tan OQN = ON/OQ = VK/VP\sin\beta = \tan\alpha\sin\beta$. So the tangent QN to the projection of the curve on the cone makes a constant angle with the radius vector OQ, showing that the projection is an equiangular spiral.

2.16. $\mathbf{r}_1 = \mathbf{r} + u\mathbf{p}$, $\mathbf{b}_1 = \pm\,\mathbf{p}$, $\mathbf{t}_1 = [\mathbf{t} + u'\mathbf{p} + u(-\,\kappa\mathbf{t} + \lambda\mathbf{b})](ds/ds_1)$. $\mathbf{t}_1.\mathbf{p} = 0$, so $u' = 0$, so u = constant, say $u = c$. This gives: $\mathbf{t}_1 = [\mathbf{t}(1 - c\kappa) + c\lambda\mathbf{b}]/[(1 - c\kappa)^2 + c^2\lambda^2]^{1/2}$; $\mathbf{t}_1' = \kappa_1\mathbf{p}_1 = [\kappa\mathbf{p}(1 - c\kappa) - c\lambda^2\mathbf{p}]/[(1 - c\kappa)^2 + c^2\lambda^2]$ + terms in $\mathbf{t}$ and $\mathbf{b}$. But $\mathbf{p}.\mathbf{p}_1 = \pm\,\mathbf{b}_1.\mathbf{p}_1 = 0$, so $\kappa(1 - c\kappa) - c\lambda^2 = 0$, giving $c(\kappa^2 + \lambda^2) = \kappa$.

2.17. $\mathbf{r}_1 = \mathbf{r} + c\mathbf{t}$, $\mathbf{t}_1 = (\mathbf{t} + \kappa c\mathbf{p})/(ds/ds_1)$, so $\mathbf{t}_1$ is parallel to the plane of $\mathbf{t}$ and $\mathbf{p}$, the osculating plane of Γ. For the locus of P_1 to be a straight line we must have $\kappa_1 = 0$. So $\mathbf{0} = \kappa_1\mathbf{p}_1 = \mathbf{t}_1' = [(\kappa\mathbf{p} + c\kappa'\mathbf{p} + c\kappa(-\,\kappa\mathbf{t} + \lambda\mathbf{b})]/[1 + c^2\kappa^2] - [c^2\kappa\kappa'(\mathbf{t} + c\kappa\mathbf{p})]/[1 + c^2\kappa^2]^2$. For this to be so the coefficients of $\mathbf{t}$, $\mathbf{p}$, $\mathbf{b}$ must all be zero; thus $(1 + c^2\kappa^2)(-\,c\kappa^2) - c^2\kappa\kappa' = 0$, $(1 + c^2\kappa^2)(\kappa + c\kappa') - c^3\kappa^2\kappa' = 0$, $c\kappa\lambda = 0$. The last condition shows that $\lambda = 0$, so the original curve must be plane (or if $\kappa = 0$ a straight line), and the other two conditions both reduce to $c\kappa' + \kappa = -\,c^2\kappa^3$, a Bernoulli type equation with solution $\kappa^{-2} + c^2 = Ae^{2s/c}$.

2.18. As in example 2.16 PP_1 = const. = a, say. Now $(\mathbf{t}.\mathbf{t}_1)' = \kappa\mathbf{p}.\mathbf{t}_1 + \mathbf{t}.\kappa_1\mathbf{p}_1(ds_1/ds) = 0$, since both $\mathbf{p}.\mathbf{t}_1$ and $\mathbf{t}.\mathbf{p}_1$ are zero. So $\mathbf{t}.\mathbf{t}_1$ is constant, and the tangents are inclined at a constant angle given by $\cos\alpha = \mathbf{t}.\mathbf{t}_1 = (1 - a\kappa)/[1 - a\kappa)^2 + a^2\lambda^2]^{1/2} = (1 - a\kappa)(ds/ds_1)$. We may also write $\sin\alpha = \pm\, a\lambda(ds/ds_1)$, so that κ, λ are now seen to satisfy the linear relation $(1 - a\kappa)\tan\alpha = \pm\, a\lambda$. As Γ may be obtained from Γ_1 in the same way, the corresponding results are found by interchanging s and s_1, altering the signs of a and of α, and by writing κ_1, λ_1 for κ, λ, so that $\cos\alpha = (1 + a\kappa_1)(ds_1/ds)$, $\sin\alpha = \pm\, a\lambda_1(ds_1/ds)$, and using $\cos^2\alpha + \sin^2\alpha = 1$, we get $(1 - a\kappa)(1 + a\kappa_1) + a^2\lambda\lambda_1 = 1$, the required result.

2.19. The vector $\mathbf{r}_1$ to a point on the involute is $\mathbf{r} + \mathbf{t}(c - s)$. If the involute lies on the plane $\mathbf{r}.\mathbf{n} = d$, then $\mathbf{r}_1.\mathbf{n} = d$, so, differentiating, $\mathbf{t}_1.\mathbf{n} = 0$, that is $[\mathbf{t} + \kappa\mathbf{p}(c - s) - \mathbf{t}].\mathbf{n} = 0$. Thus $\mathbf{p}.\mathbf{n} = 0$, $\kappa\mathbf{p}.\mathbf{n} = 0$, and integrating, $\mathbf{t}.\mathbf{n}$ = const., so the original curve is a helix.

2.20. If P_1 lies on Γ_1, an evolute of Γ, since $\mathbf{t}_1$ is normal to Γ $\mathbf{r}_1 = \mathbf{r} + \mu\mathbf{p} + \nu\mathbf{b}$, for some μ,ν. Hence $\mathbf{t}_1 = [\mathbf{t} + \mu(-\,\kappa\mathbf{t} + \lambda\mathbf{b}) - \nu\lambda\mathbf{b} + \mu'\mathbf{p} + \nu'\mathbf{b}](ds/ds_1)$, and this is parallel to $\mu\mathbf{p} + \nu\mathbf{b}$ in the normal plane at P to Γ. So $1 - \mu\kappa = 0$, $\mu = 1/\kappa$, $\mathbf{r}_1 = \mathbf{r} + \kappa^{-1}\mathbf{p} + \nu\mathbf{b}$. So P_1 lies on a line parallel to $\mathbf{b}$ through C, the centre of curvature of Γ at P.

Chapter Three

3.1. $ds^2 = Edu^2 + Gdv^2$. $E^{1/2}du/G^{1/2}dv = \tan\alpha$.

3.2. $\mathbf{r} = \mathbf{i}a(u + v) + \mathbf{j}b(u - v) + 2\mathbf{k}cuv$. $bx + ay$ = const., $bx - ay$ = const., $c(b^2x^2 - a^2y^2) = 2a^2b^2z$.

3.3. $E = a^2\sin\phi$, $F = 0$, $G = a^2$. $ds^2 = a^2(d\phi^2 + \sin^2\phi\, d\theta^2)$. (i) $a\pi$; (ii) $2a\pi \sin\beta$; (iii) $ds = a(1 + \sin^2\phi\sec^2\phi)^{1/2}\,d\phi = a\sec\phi\, d\phi$,

$$s = \Big[a\log(\sec\phi + \tan\phi)\Big]_0^{\pi/4} = a\log(\sqrt{2} + 1).$$

$$A = \int_0^{\pi}\int_0^{2\pi} a^2\sin\phi\, d\theta\, d\phi = 2\pi a^2\int_0^{\pi}\sin\phi\, d\phi = 4\pi a^2\,.$$

3.4. A surface generated by a horizontal line passing through the axis of z as the line moves along the axis, turning about it always in the same sense.

$d\mathbf{r} = (\mathbf{i}\cos\theta + \mathbf{j}\sin\theta)du + (-\mathbf{i}u\sin\theta + \mathbf{j}u\cos\theta + \mathbf{k}c)d\theta$, $d\mathbf{r}_1.d\mathbf{r}_2 = du_1du_2 + (u^2 + c^2)d\theta_1d\theta_2 = 0$ since $du_1/d\theta_1$, $du_2/d\theta_2$ are the roots of $(du/d\theta)^2 - (u^2 + c^2) = 0$.

3.5. $ds^2 = (1 + \tan^2 u)du^2 - 2\tan u\, du d\theta + (u^2 + 1)d\theta^2$. For curves $\theta = \lambda$, $ds = \sec u\, du$; $s = [\log(\sec u + \tan u)]_a^b$, independent of λ.

3.6. Taking lines of curvature as parametric curves,

$$\kappa_N = (Ldu^2 + Ndv^2)/(Edu^2 + Gdv^2),\quad E^{1/2}\cos\alpha\, du = G^{1/2}\sin\alpha\, dv,$$

$$\kappa_N = (L/E)\sin^2\alpha + (N/G)\cos^2 a = K_2\sin^2\alpha + K_1\cos^2 a.$$

3.7. For perpendicular directions, using 3.6, the sum of the curvatures is $(K_1\cos^2\alpha + K_2\sin^2\alpha) + (K_1\sin^2\alpha + K_2\cos^2\alpha) = K_1 + K_2$.

3.8. u is the distance of a point on the surface from the axis of symmetry. $E = 1 + [f'(u)]^2$, $F = 0$, $G = u^2$. $s = \int[(1 + [f'(u)]^2)du^2 + u^2d\theta^2]^{1/2}$ with $du = g(u,\theta)d\theta$, where $g(u,\theta)$ is determined by Γ, given by $h(u,\theta) = 0$. $\mathbf{r}_{uu} = (0,0,f''(u))$, $\mathbf{r}_{u\theta} = (-\sin\theta, \cos\theta, 0)$, $\mathbf{r}_{\theta\theta} = (-u\cos\theta, -u\sin\theta, 0)$. $\mathbf{r}_u \wedge \mathbf{r}_\theta = (-uf'(u)\cos\theta,\ -uf'(u)\sin\theta, u)$, $LW = uf''(u)$, $MW = 0$, $NW = u^2f'(u)$. Lines of curvature are

$$\begin{vmatrix} d\theta^2 & -d\theta du & du^2 \\ 1 + [f'(u)]^2 & 0 & u^2 \\ uf''(u) & 0 & u^2f'(u) \end{vmatrix} = 0,$$

i.e. $dud\theta = 0$, i.e. $u = \text{const.}$, $\theta = \text{const.}$ The first set gives circular sections perpendicular to the axis of symmetry, and the second set gives the meridians.

3.9. A surface formed by rotating a catenary about its directrix. $\mathbf{r} = \mathbf{i}u\cos\theta + \mathbf{j}u\sin\theta + \mathbf{k}\, c\cosh^{-1}(u/c)$, $E = u^2/(u^2 - c^2)$, $F = 0$, $G = u^2$. $LW = -u^2c/(u^2 - c^2)^{3/2}$, $MW = 0$, $NW = u^2c/(u^2 - c^2)^{1/2}$. Lines of curvature are $dud\theta = 0$. Asymptotic lines are $du^2 - (u^2 - c^2)d\theta^2 = 0$. Perpendicularity as in 3.4 since $d\mathbf{r}_1.d\mathbf{r}_2 = du_1du_2 + (u^2 - c^2)d\theta_1d\theta_2$.

3.10. Using the lines of curvature as parametric curves $I = (d\mathbf{r})^2 = (d\mathbf{r})_u^2 + (d\mathbf{r})_v^2$, $II = -d\mathbf{r}.d\mathbf{e} = -\{(d\mathbf{r})_u + (d\mathbf{r})_v\}.\{(d\mathbf{e})_u + (d\mathbf{e})_v\} = \{(d\mathbf{r})_u + (d\mathbf{r})_v\}.\{+K_1(d\mathbf{r})_u + K_2(d\mathbf{r})_v\} = K_1(d\mathbf{r})_u^2 + K_2(d\mathbf{r})_v^2$, $III = (d\mathbf{e})^2 = (d\mathbf{e})_u^2 + (d\mathbf{e})_v^2 = K_1^2(d\mathbf{r})_u^2 + K_2^2(d\mathbf{r})_v^2$, so

$$\begin{vmatrix} \mathrm{I} & 1 & 1 \\ \mathrm{II} & K_1 & K_2 \\ \mathrm{III} & K_1 & K_2 \end{vmatrix} = 0, \text{ i.e. } K_1K_2I - (K_1 + K_2)II + III = 0.$$

3.11. For asymptotic lines $II = 0$, so $KI + III = 0$; also $\mathbf{b} = \mathbf{e}$, so $\mathbf{b}' = -\lambda\mathbf{p}$ becomes $-\lambda\mathbf{p} = d\mathbf{e}/ds$, so $\lambda^2 = (d\mathbf{e})^2/(ds)^2 = III/I = -K$. Thus $\lambda = \pm\sqrt{(-K)}$.

3.12. $\mathbf{R}_s = (1 - u\kappa)\mathbf{t} + u\lambda\mathbf{b}$, $\mathbf{R}_u = \mathbf{p}$, $\mathbf{R}_{ss} = + \kappa\mathbf{p} + u(-\kappa'\mathbf{t} + \lambda'\mathbf{b}) - u\mathbf{p}(\kappa^2 + \lambda^2)$, $\mathbf{R}_{su} = -\kappa\mathbf{t} + \lambda\mathbf{b}$, $\mathbf{R}_{uu} = 0$, $\mathbf{R}_s \wedge \mathbf{R}_u = (1 - u\kappa)\mathbf{b} - u\lambda\mathbf{t}$, $LW = u^2\lambda\kappa' + (1 - u\kappa)\lambda' u$, $MW = \lambda$, $NW = 0$. Equation giving curvatures K_1, K_2 is $K_N(EG - F^2) - K_N(EN + 2FM + GL) + (LN - M^2) = 0$, $K_N^2(1 - 2u\kappa + u^2[\kappa^2 + \lambda^2]) - K_N(u^2[\lambda\kappa' - \kappa\lambda'] + \lambda' u)/W - \lambda^2/W^2 = 0$.

3.13. Using lines of curvature, $dudv = 0$, as parametric curves, the asymptotic lines are $Ldu^2 + Ndv^2 = 0$, or $du/dv = \pm\sqrt{(-N/L)}$, showing that their directions are bisected by the lines of curvature.

3.14. S cuts π at a constant angle. Any curve on π is a line of curvature since both de/ds and κ are zero, so that $de/ds + \kappa\mathbf{t} = 0$ holds. Thus the curve is a line of curvature on S.

3.15. $E = 1$, $F = 0$, $G = \lambda^2$.

$$K\lambda^4 = \begin{vmatrix} 1 & 0 & -\lambda\lambda_u \\ 0 & \lambda^2 & \lambda\lambda_v \\ 0 & 0 & -\lambda_u^2 - \lambda\lambda_{uu} \end{vmatrix} - \begin{vmatrix} 1 & 0 & 0 \\ 0 & \lambda^2 & \lambda\lambda_u \\ 0 & \lambda\lambda_u & 0 \end{vmatrix}$$

$= \lambda^2(-\lambda_u^2 - \lambda\lambda_{uu}) + \lambda^2\lambda_u^2$. So

$$K = -\frac{1}{\lambda}\frac{\partial^2\lambda}{\partial u^2}$$

which, for a sphere, gives

$$K = -\frac{1}{a\sin\phi}\,\frac{\partial^2(a\sin\phi)}{a^2\partial\phi^2} = \frac{1}{a^2}.$$

3.16. $\mathbf{r}_u = (\cos\theta, \sin\theta, 0)$, $\mathbf{r}_\theta = (-u\sin\theta, u\cos\theta, c)$, $\mathbf{r}_{uu} = (0,0,0)$, $\mathbf{r}_{u\theta} = (-\sin\theta, \cos\theta, 0)$, $\mathbf{r}_{\theta\theta} = (-u\cos\theta, -u\sin\theta, 0)$, $\mathbf{r}_u \wedge \mathbf{r}_\theta = (c\sin\theta, -c\cos\theta, u)$; $LW = 0$, $MW = -c$, $NW = 0$, so the asymptotic lines are $dud\theta = 0$, the parametric curves. The lines of curvature are

$$\begin{vmatrix} d\theta^2 & -d\theta du & du^2 \\ 1 & 0 & u^2 + c^2 \\ 0 & -c & 0 \end{vmatrix} = 0,$$

that is $du^2 - (u^2 + c^2)d\theta^2 = 0$,
and the principal curvatures, K_N, are given by $K_N^2(u^2 + c^2)^2 - c^2 = 0$, so they are equal in magnitude, opposite in sign, and independent of θ.

3.17. Direction ratios for the normal are f_x, f_y, f_z, so at (α,β,γ) they are $K_1\alpha$, $K_2\beta$, -1. Any point on the normal is $(\alpha + \sigma K_1\alpha, \beta + \sigma K_2\beta, \gamma - \sigma)$, which lies on $x = 0$, $y = 0$ if $\alpha = 0$, $1 + \sigma K_2 = 0$, or $1 + \sigma K_1 = 0$, $\beta = 0$, that is if (α,β,γ) lies on one or other of the lines of curvature through O.

3.18. $\mathbf{r}_u = (a,b,2cv)$, $\mathbf{r}_v = (a, -b, 2cu)$, $\mathbf{r}_{uu} = (0,0,0)$, $\mathbf{r}_{uv} = (0,0,2c)$, $\mathbf{r}_{vv} = (0,0,0)$, $\mathbf{r}_u \wedge \mathbf{r}_v = \{2bc(u + v), 2ca(v - u), -2ab\}$, $LW = 0$, $MW = -4abc$, $NW = 0$. The asymptotic lines are $dudv = 0$, the parametric lines. The lines of curvature are $dv^2(a^2 + b^2 + 4c^2u^2) - du^2(a^2 + b^2 + 4c^2v^2) = 0$, that is

$$\sinh^{-1}\frac{2cu}{\sqrt{(a^2 + b^2)}} \pm \sinh^{-1}\frac{2cv}{\sqrt{(a^2 + b^2)}} = \text{constant}.$$

3.19. $\mathbf{r}_x = \mathbf{i} + \mathbf{k}p$, $\mathbf{r}_y = \mathbf{j} + \mathbf{k}q$, $\mathbf{r}_{xx} = \mathbf{k}r$, $\mathbf{r}_{xy} = \mathbf{k}s$, $\mathbf{r}_{yy} = \mathbf{k}t$, $\mathbf{r}_x \wedge \mathbf{r}_y = -\mathbf{i}p - \mathbf{j}q + \mathbf{k}$, $E = 1 + p^2$, $F = pq$, $G = 1 + q^2$, $LW = r$, $MW = s$, $NW = t$, $W = \sqrt{(1 + p^2 + q^2)}$, $\mathbf{e} = (-\mathbf{i}p - \mathbf{j}q + \mathbf{k})/W$, $K = (rt - s^2)/(1 + p^2 + q^2)^2$, and $rt = s^2$ for a developable. The tangent plane is $(X - x)p + (Y - y)q - (Z - z) = 0$, and the direction cosines of normals to it are $(p, q, -1)/\sqrt{(1 + p^2 + q^2)}$. For this to be a function of a single parameter there must be a functional relation between p and q, say $F(p,q) = 0$, in which case

$$\frac{\partial F}{\partial p}\frac{\partial p}{\partial x} + \frac{\partial F}{\partial q}\frac{\partial q}{\partial x} = 0,$$

and

$$\frac{\partial F}{\partial p}\frac{\partial p}{\partial y} + \frac{\partial F}{\partial q}\frac{\partial q}{\partial y} = 0,$$

that is $F_p r + F_q s = 0$, $F_p s + F_q t = 0$, whence

$$\begin{vmatrix} r & s \\ s & t \end{vmatrix} = 0, \text{ that is } rt = s^2.$$

3.20. $\mathbf{R}_s = \mathbf{t}(s) + u\kappa\mathbf{p}(s)$, $\mathbf{R}_u = \mathbf{t}(s)$, $E = 1 + u^2\kappa^2$, $F = 1$, $G = 1$, $\mathbf{R}_{ss} = \kappa\mathbf{p} + u\kappa'\mathbf{p} + u\kappa(-\kappa\mathbf{t} + \lambda\mathbf{b})$, $R_{su} = \kappa\mathbf{p}(s)$, $\mathbf{R}_{uu} = 0$, $\mathbf{R}_s \wedge \mathbf{R}_u = -u\kappa\mathbf{b}$, $LW = -u^2\kappa^2\lambda$, $MW = 0$, $NW = 0$, so the lines of curvature are

$$\begin{vmatrix} du^2 & -duds & ds^2 \\ 1 + u^2\kappa^2 & 1 & 1 \\ -u^2\kappa^2\lambda & 0 & 0 \end{vmatrix} = 0,$$

that is $ds(ds + du) = 0$, that is $s =$ const., $s + u =$ const. The asymptotic lines are $ds^2 = 0$, that is $s =$ const., the generators.

3.21. $LN - M^2 = (\mathbf{r}_u.\mathbf{e}_u)(\mathbf{r}_v.\mathbf{e}_v) - (\mathbf{r}_u.\mathbf{e}_v)(\mathbf{r}_v.\mathbf{e}_u) = (\mathbf{r}_u \wedge \mathbf{r}_v).(\mathbf{e}_u \wedge \mathbf{e}_v) = \mathbf{e}W.(\mathbf{e}_u \wedge \mathbf{e}_v) = W[\mathbf{e},\mathbf{e}_u,\mathbf{e}_v]$. Now $\mathbf{e}_u$ is parallel to $\mathbf{e}_v$, so $LN - M^2 = 0$. Choose the asymptotic lines as one set of parametric curves, that is $(du\sqrt{L} + dv\sqrt{N})^2 = 0$, and choose a new set of parameters with $du' = du\sqrt{L} + dv\sqrt{N}$. Then $M' = N' = 0$, so $r_{u'}.\mathbf{e}_{v'} = \mathbf{r}_{v'}.\mathbf{e}_{v'} = 0$, and $\mathbf{e}_{v'} = \mathbf{0}$, and $\mathbf{e}_{v'}$ depends on one parameter only, and the surface is developable.

3.22. The equation of the normal plane is $(\mathbf{R} - \mathbf{r}).\mathbf{t} = 0$, so $-\mathbf{t}.\mathbf{t} + (\mathbf{R} - \mathbf{r}).\kappa\mathbf{p} = 0$, $-\mathbf{t}.\kappa\mathbf{p} + (\mathbf{R} - \mathbf{r}).\kappa'\mathbf{p} + (\mathbf{R} - \mathbf{r}).\kappa(-\kappa\mathbf{t} + \lambda\mathbf{b}) = 0$. So we have $(\mathbf{R} - \mathbf{r}).\mathbf{t} = 0$, $(\mathbf{R} - \mathbf{r}).\mathbf{p} = 1/\kappa$, $(\mathbf{R} - \mathbf{r}).\mathbf{b} = (\mathbf{R} - \mathbf{r}).(\kappa^2\mathbf{t} - \kappa'\mathbf{p})/\kappa\lambda = -\kappa'/\kappa^2\lambda$. So $(\mathbf{R} - \mathbf{r}) = [0(\mathbf{p} \wedge \mathbf{b}) + \kappa^{-1}(\mathbf{b} \wedge \mathbf{t}) - \kappa'(\mathbf{t} \wedge \mathbf{p})/\kappa^2\lambda]/[\mathbf{p} \wedge \mathbf{b}, \mathbf{b} \wedge \mathbf{t}, \mathbf{t} \wedge \mathbf{p}] = \kappa^{-1}\mathbf{p} - \kappa'\mathbf{b}/\kappa^2\lambda$, the vector from the point to the centre of spherical curvature.

3.23. For a geodesic $\mathbf{p} = -\mathbf{e}$. Now $\mathbf{b}' = -\lambda\mathbf{p}$, so $\lambda\mathbf{e} = (\mathbf{t} \wedge \mathbf{p})' = -(\mathbf{t} \wedge \mathbf{e})' = \mathbf{e}' \wedge \mathbf{t} + \mathbf{e} \wedge \mathbf{t}' = \mathbf{e}' \wedge \mathbf{r}' + \mathbf{e} \wedge \kappa\mathbf{p} = \mathbf{e}' \wedge \mathbf{r}'$. So $\lambda\mathbf{e}.\mathbf{e} = \lambda = \mathbf{e}.(\mathbf{e}' \wedge \mathbf{r}') = [\mathbf{e},\mathbf{e}',\mathbf{r}']$.

3.24. $\boldsymbol{\kappa}_G = \pm\, \mathbf{e} \wedge \boldsymbol{\kappa} = \pm\, \mathbf{e} \wedge \mathbf{t}' = \pm\, \mathbf{e} \wedge \mathbf{r}''$. But $\boldsymbol{\kappa}_G$ is parallel to $\mathbf{t}$, which is $\mathbf{r}'$, so $\kappa_G = \pm\, \mathbf{r}'.\boldsymbol{\kappa}_G = \pm\, \mathbf{r}'.(\mathbf{e} \wedge \mathbf{r}'') = \pm\, [\mathbf{e},\mathbf{r}',\mathbf{r}'']$.

3.25. $\mathbf{R} = \mathbf{r} + u\mathbf{b}$, $\mathbf{R}_s = \mathbf{t} - u\lambda\mathbf{p}$, $\mathbf{R}_u = \mathbf{b}$, $\mathbf{e}$ is parallel to $R_s \wedge \mathbf{R}_u = -u\lambda\mathbf{t} - \mathbf{p}$. So $\mathbf{e}.\mathbf{b} = 0$, and Γ is a geodesic.

Index

INDEX

Chemistry

THE SCEPTICAL CHYMIST: THE CLASSIC 1661 TEXT, Robert Boyle. Boyle defines the term "element," asserting that all natural phenomena can be explained by the motion and organization of primary particles. 1911 ed. viii+232pp. 5⅜ x 8½.
0-486-42825-7

RADIOACTIVE SUBSTANCES, Marie Curie. Here is the celebrated scientist's doctoral thesis, the prelude to her receipt of the 1903 Nobel Prize. Curie discusses establishing atomic character of radioactivity found in compounds of uranium and thorium; extraction from pitchblende of polonium and radium; isolation of pure radium chloride; determination of atomic weight of radium; plus electric, photographic, luminous, heat, color effects of radioactivity. ii+94pp. 5⅜ x 8½. 0-486-42550-9

CHEMICAL MAGIC, Leonard A. Ford. Second Edition, Revised by E. Winston Grundmeier. Over 100 unusual stunts demonstrating cold fire, dust explosions, much more. Text explains scientific principles and stresses safety precautions. 128pp. 5⅜ x 8½. 0-486-67628-5

THE DEVELOPMENT OF MODERN CHEMISTRY, Aaron J. Ihde. Authoritative history of chemistry from ancient Greek theory to 20th-century innovation. Covers major chemists and their discoveries. 209 illustrations. 14 tables. Bibliographies. Indices. Appendices. 851pp. 5⅜ x 8½. 0-486-64235-6

CATALYSIS IN CHEMISTRY AND ENZYMOLOGY, William P. Jencks. Exceptionally clear coverage of mechanisms for catalysis, forces in aqueous solution, carbonyl- and acyl-group reactions, practical kinetics, more. 864pp. 5⅜ x 8½.
0-486-65460-5

ELEMENTS OF CHEMISTRY, Antoine Lavoisier. Monumental classic by founder of modern chemistry in remarkable reprint of rare 1790 Kerr translation. A must for every student of chemistry or the history of science. 539pp. 5⅜ x 8½. 0-486-64624-6

THE HISTORICAL BACKGROUND OF CHEMISTRY, Henry M. Leicester. Evolution of ideas, not individual biography. Concentrates on formulation of a coherent set of chemical laws. 260pp. 5⅜ x 8½. 0-486-61053-5

A SHORT HISTORY OF CHEMISTRY, J. R. Partington. Classic exposition explores origins of chemistry, alchemy, early medical chemistry, nature of atmosphere, theory of valency, laws and structure of atomic theory, much more. 428pp. 5⅜ x 8½. (Available in U.S. only.) 0-486-65977-1

GENERAL CHEMISTRY, Linus Pauling. Revised 3rd edition of classic first-year text by Nobel laureate. Atomic and molecular structure, quantum mechanics, statistical mechanics, thermodynamics correlated with descriptive chemistry. Problems. 992pp. 5⅜ x 8½. 0-486-65622-5

FROM ALCHEMY TO CHEMISTRY, John Read. Broad, humanistic treatment focuses on great figures of chemistry and ideas that revolutionized the science. 50 illustrations. 240pp. 5⅜ x 8½. 0-486-28690-8

Engineering

DE RE METALLICA, Georgius Agricola. The famous Hoover translation of greatest treatise on technological chemistry, engineering, geology, mining of early modern times (1556). All 289 original woodcuts. 638pp. 6¾ x 11. 0-486-60006-8

FUNDAMENTALS OF ASTRODYNAMICS, Roger Bate et al. Modern approach developed by U.S. Air Force Academy. Designed as a first course. Problems, exercises. Numerous illustrations. 455pp. 5⅜ x 8½. 0-486-60061-0

DYNAMICS OF FLUIDS IN POROUS MEDIA, Jacob Bear. For advanced students of ground water hydrology, soil mechanics and physics, drainage and irrigation engineering and more. 335 illustrations. Exercises, with answers. 784pp. 6⅛ x 9¼. 0-486-65675-6

THEORY OF VISCOELASTICITY (Second Edition), Richard M. Christensen. Complete consistent description of the linear theory of the viscoelastic behavior of materials. Problem-solving techniques discussed. 1982 edition. 29 figures. xiv+364pp. 6⅛ x 9¼. 0-486-42880-X

MECHANICS, J. P. Den Hartog. A classic introductory text or refresher. Hundreds of applications and design problems illuminate fundamentals of trusses, loaded beams and cables, etc. 334 answered problems. 462pp. 5⅜ x 8½. 0-486-60754-2

MECHANICAL VIBRATIONS, J. P. Den Hartog. Classic textbook offers lucid explanations and illustrative models, applying theories of vibrations to a variety of practical industrial engineering problems. Numerous figures. 233 problems, solutions. Appendix. Index. Preface. 436pp. 5⅜ x 8½. 0-486-64785-4

STRENGTH OF MATERIALS, J. P. Den Hartog. Full, clear treatment of basic material (tension, torsion, bending, etc.) plus advanced material on engineering methods, applications. 350 answered problems. 323pp. 5⅜ x 8½. 0-486-60755-0

A HISTORY OF MECHANICS, René Dugas. Monumental study of mechanical principles from antiquity to quantum mechanics. Contributions of ancient Greeks, Galileo, Leonardo, Kepler, Lagrange, many others. 671pp. 5⅜ x 8½. 0-486-65632-2

STABILITY THEORY AND ITS APPLICATIONS TO STRUCTURAL MECHANICS, Clive L. Dym. Self-contained text focuses on Koiter postbuckling analyses, with mathematical notions of stability of motion. Basing minimum energy principles for static stability upon dynamic concepts of stability of motion, it develops asymptotic buckling and postbuckling analyses from potential energy considerations, with applications to columns, plates, and arches. 1974 ed. 208pp. 5⅜ x 8½. 0-486-42541-X

METAL FATIGUE, N. E. Frost, K. J. Marsh, and L. P. Pook. Definitive, clearly written, and well-illustrated volume addresses all aspects of the subject, from the historical development of understanding metal fatigue to vital concepts of the cyclic stress that causes a crack to grow. Includes 7 appendixes. 544pp. 5⅜ x 8½. 0-486-40927-9

Mathematics

FUNCTIONAL ANALYSIS (Second Corrected Edition), George Bachman and Lawrence Narici. Excellent treatment of subject geared toward students with background in linear algebra, advanced calculus, physics and engineering. Text covers introduction to inner-product spaces, normed, metric spaces, and topological spaces; complete orthonormal sets, the Hahn-Banach Theorem and its consequences, and many other related subjects. 1966 ed. 544pp. 6⅛ x 9¼. 0-486-40251-7

ASYMPTOTIC EXPANSIONS OF INTEGRALS, Norman Bleistein & Richard A. Handelsman. Best introduction to important field with applications in a variety of scientific disciplines. New preface. Problems. Diagrams. Tables. Bibliography. Index. 448pp. 5⅜ x 8½. 0-486-65082-0

VECTOR AND TENSOR ANALYSIS WITH APPLICATIONS, A. I. Borisenko and I. E. Tarapov. Concise introduction. Worked-out problems, solutions, exercises. 257pp. 5⅝ x 8¼. 0-486-63833-2

AN INTRODUCTION TO ORDINARY DIFFERENTIAL EQUATIONS, Earl A. Coddington. A thorough and systematic first course in elementary differential equations for undergraduates in mathematics and science, with many exercises and problems (with answers). Index. 304pp. 5⅜ x 8½. 0-486-65942-9

FOURIER SERIES AND ORTHOGONAL FUNCTIONS, Harry F. Davis. An incisive text combining theory and practical example to introduce Fourier series, orthogonal functions and applications of the Fourier method to boundary-value problems. 570 exercises. Answers and notes. 416pp. 5⅜ x 8½. 0-486-65973-9

COMPUTABILITY AND UNSOLVABILITY, Martin Davis. Classic graduate-level introduction to theory of computability, usually referred to as theory of recurrent functions. New preface and appendix. 288pp. 5⅜ x 8½. 0-486-61471-9

ASYMPTOTIC METHODS IN ANALYSIS, N. G. de Bruijn. An inexpensive, comprehensive guide to asymptotic methods–the pioneering work that teaches by explaining worked examples in detail. Index. 224pp. 5⅜ x 8½ 0-486-64221-6

APPLIED COMPLEX VARIABLES, John W. Dettman. Step-by-step coverage of fundamentals of analytic function theory–plus lucid exposition of five important applications: Potential Theory; Ordinary Differential Equations; Fourier Transforms; Laplace Transforms; Asymptotic Expansions. 66 figures. Exercises at chapter ends. 512pp. 5⅜ x 8½. 0-486-64670-X

INTRODUCTION TO LINEAR ALGEBRA AND DIFFERENTIAL EQUATIONS, John W. Dettman. Excellent text covers complex numbers, determinants, orthonormal bases, Laplace transforms, much more. Exercises with solutions. Undergraduate level. 416pp. 5⅜ x 8½. 0-486-65191-6

RIEMANN'S ZETA FUNCTION, H. M. Edwards. Superb, high-level study of landmark 1859 publication entitled "On the Number of Primes Less Than a Given Magnitude" traces developments in mathematical theory that it inspired. xiv+315pp. 5⅜ x 8½. 0-486-41740-9

CATALOG OF DOVER BOOKS

TENSOR CALCULUS, J.L. Synge and A. Schild. Widely used introductory text covers spaces and tensors, basic operations in Riemannian space, non-Riemannian spaces, etc. 324pp. 5⅜ x 8¼. 0-486-63612-7

ORDINARY DIFFERENTIAL EQUATIONS, Morris Tenenbaum and Harry Pollard. Exhaustive survey of ordinary differential equations for undergraduates in mathematics, engineering, science. Thorough analysis of theorems. Diagrams. Bibliography. Index. 818pp. 5⅜ x 8½. 0-486-64940-7

INTEGRAL EQUATIONS, F. G. Tricomi. Authoritative, well-written treatment of extremely useful mathematical tool with wide applications. Volterra Equations, Fredholm Equations, much more. Advanced undergraduate to graduate level. Exercises. Bibliography. 238pp. 5⅜ x 8½. 0-486-64828-1

FOURIER SERIES, Georgi P. Tolstov. Translated by Richard A. Silverman. A valuable addition to the literature on the subject, moving clearly from subject to subject and theorem to theorem. 107 problems, answers. 336pp. 5⅜ x 8½. 0-486-63317-9

INTRODUCTION TO MATHEMATICAL THINKING, Friedrich Waismann. Examinations of arithmetic, geometry, and theory of integers; rational and natural numbers; complete induction; limit and point of accumulation; remarkable curves; complex and hypercomplex numbers, more. 1959 ed. 27 figures. xii+260pp. 5⅜ x 8½. 0-486-63317-9

POPULAR LECTURES ON MATHEMATICAL LOGIC, Hao Wang. Noted logician's lucid treatment of historical developments, set theory, model theory, recursion theory and constructivism, proof theory, more. 3 appendixes. Bibliography. 1981 edition. ix + 283pp. 5⅜ x 8½. 0-486-67632-3

CALCULUS OF VARIATIONS, Robert Weinstock. Basic introduction covering isoperimetric problems, theory of elasticity, quantum mechanics, electrostatics, etc. Exercises throughout. 326pp. 5⅜ x 8½. 0-486-63069-2

THE CONTINUUM: A CRITICAL EXAMINATION OF THE FOUNDATION OF ANALYSIS, Hermann Weyl. Classic of 20th-century foundational research deals with the conceptual problem posed by the continuum. 156pp. 5⅜ x 8½. 0-486-67982-9

CHALLENGING MATHEMATICAL PROBLEMS WITH ELEMENTARY SOLUTIONS, A. M. Yaglom and I. M. Yaglom. Over 170 challenging problems on probability theory, combinatorial analysis, points and lines, topology, convex polygons, many other topics. Solutions. Total of 445pp. 5⅜ x 8½. Two-vol. set.
Vol. I: 0-486-65536-9 Vol. II: 0-486-65537-7

Paperbound unless otherwise indicated. Available at your book dealer, online at **www.doverpublications.com**, or by writing to Dept. GI, Dover Publications, Inc., 31 East 2nd Street, Mineola, NY 11501. For current price information or for free catalogues (please indicate field of interest), write to Dover Publications or log on to **www.doverpublications.com** and see every Dover book in print. Dover publishes more than 500 books each year on science, elementary and advanced mathematics, biology, music, art, literary history, social sciences, and other areas.